Jürgen Schultz
The Ecozones of the World
The Ecological Divisions of the Geosphere

Jürgen Schultz

The Ecozones
of the World

The Ecological Divisions
of the Geosphere

Translated into English by Bridget Ahnert

Second Edition

with 144 Figures, 22 Tables, and 5 Boxes

 Springer

Author

Jürgen Schultz
Geographisches Institut
Rheinisch-Westfälische
Technische Hochschule Aachen (RWTH)
Templergraben 55
52056 Aachen
Germany

Translator

Bridget Ahnert
Karl-Christ-Str. 15a
69118 Heidelberg

This book was originally published in German under the title "Die Ökozonen der Erde", 3. Auflage.
© 2002 by Eugen Ulmer GmbH & Co., Stuttgart, Germany
1st English Edition Springer-Verlag, 1995

ISBN-10 3-540-20014-2 Springer Berlin Heidelberg New York
ISBN-13 978-3-540-20014-7 Springer Berlin Heidelberg New York

Library of Congress Control Number: 2005924522

Springer is a part of Springer Science+Business Media

springeronline.com

© Springer-Verlag Berlin Heidelberg 2005
Printed in the Netherlands

The use of designations, trademarks, etc. in this publication does not imply, even in the absence of a specific statement, that such names are exempt from the relevant protective laws and regulations and therefore free for general use.

Typesetting: LE-TeX Jelonek, Schmidt & Vöckler GbR, Leipzig
Cover design: E. Kirchner, Heidelberg
Production: A. Oelschläger
Printed on acid-free paper   30/2132/AO   5 4 3 2 1 0

Preface

Nine terrestrial ecozones are distinguished and described in separate chapters. Each of these regional chapters is subdivided by the same headings, i.e. distribution, climate, relief and drainage, soil, vegetation and animals, and land use. The contents of the corresponding subdivisions have similar organizations and use consistent terminology and units of measurement. This should help the reader greatly in finding and comparing information which especially interests him; e.g. he can easily find which soil units are characteristic for each of the ecozones or how ecozones differ with respect to soil units. Other features and processes handled in this way include weathering of rocks, erosion and sedimentation processes, soil formation, solar radiation, growing seasons, moisture regime, vegetation structure and dynamics, nutrient cycling, energy fluxes, ecosystem models, agricultural use and potential.

The regional section which presents the nine individual ecozones is preceded by a general section in which numerous ecological termini are introduced and comparing overviews of selected characteristics are described.

Soil types are classified and named in accordance with the FAO-UNESCO classification system, which has been utilized in preparing a unique detailed soil map of the entire earth (and an even more detailed one covering the European Economic Community). With this map, internationally applicable soil terminology has been made available for the first time. The terminology may sound somewhat foreign to unaccustomed ears, but this will not be the case for long. The reader should not let himself be put off by foreign-sounding soil names – he will have to learn them sooner or later anyway!

This book was conceived mainly to serve students of geography. I would be pleased if others also find interest in it – students of biology, agricultural sciences or forestry, and geography or biology teachers striving to further their knowledge, and all those interested in ecology and geography who would like more information on the specific characteristics of the major regions of the earth, if only for the purpose of preparing themselves for a journey to another part of our planet. During my stays in foreign regions, I have at times wished for a book which – as is the intended purpose of the present volume – could have provided in concentrated form a summary and explanation of the most significant characteristics of the major regions of the earth and their interactions with one another.

The theme of the book in hand is (geo)ecology. Not however in the current prevailing sense, whereby the (destructive) environmental effects of man and the protection of the environment from these influences have taken centre stage. It is much more concerned with furnishing us with information about the nature of our environment, its form and the materials of its composite parts, as well as the interaction between them. It goes without saying that hope will be born for a greater understanding for the environment as well as its preservation.

Jürgen Schultz
Aachen, August 2004

Contents

General: the treatment of the ecozones and global overviews
of selected characteristics

Regional section: The individual ecozones

General: the treatment of the ecozones and global overviews of selected characteristics

An ecozone is a large region on the terrestrial landmass of the world where physical factors such as climate, soils, landforms and rocks interact to form an original environment in which a mix of plant life grows and provides a habitat for animal life. Because it is on the surface of the earth, the cover of vegetation is the most visual expression in each ecozone of its ecology, whether it is still the natural cover, or the result of man's use of the land, or even absent altogether in a desert.

The spatial distribution of a forest and its species mix, the types of shrubs or grasslands, peat bogs or lichens and the animal life that inhabits them are all a function of the climate in which they develop and the soils in which they

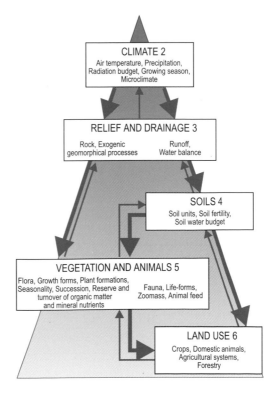

Fig. 0.1. Hierarchy of the chief characteristics in the ecozones to clarify the structure of the contents in this book. The chapter sequence is in keeping with the hierarchical interaction of these characteristics The figures (for example Climate 2) relate to the relevant chapters in the general part that is to say to the sub-chapters in the regional (the first equivalent decimal place: in the examples on climate to the sub-chapters 8.2, 9.2. 10.2 etc). In each of the boxes for the chief characteristics there is a selection of separate components to which further information is given.

grow. The climate and the rocks have played a major role in determining the quality of the soils, although they too have been influenced in their formation by the plant growth that became established on them. The original cover of vegetation in large parts of the world has been cleared for cultivation or the forest cut for timber, firewood or other uses. But the pattern of agriculture, the crops grown, whether they are grains or tree crops, such as citrus, and the animals grazed are to a large extent the optimal expression of what is possible given the physical environment of a particular ecozone.

Modern agriculture has extended the limits of many types of farming but it is the fundamental framework set by the physical factors and plant cover that determine the land use patterns that have evolved.

Nine ecozones are defined. They occur in bands, often fragmented because of the distribution of the continents and oceans, from the poles to the equator. Nearly all are present in both the Northern and Southern Hemispheres. The boundaries of the ecozones are rarely sharply drawn, although large scale temperature and precipitation data are often used because climatic thresholds define the limits of different types of plant growth and the growing season. The occurrence of frosts, periods of low temperatures, dry seasons or the coincidence of suitable temperatures for growth and moisture supply are examples of boundary forming climatic parameters for various types of vegetation and therefore also ecozones.

Chapter 1 describes the spatial distribution and size of the ecozones. Chapter 2 to 6 provide a general discussion of the role of climate, relief and drainage, soils vegetation, animal life and land use. In chapters 7 to 15 each of these aspects is described in more detail so that the interrelationships between them can be explained and the character of the ecozone understood.

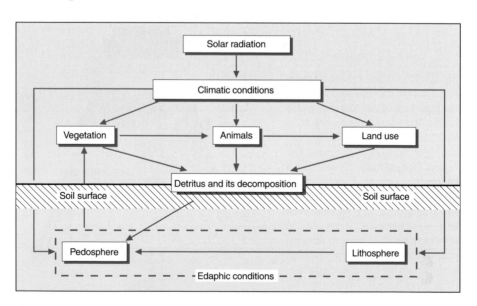

Fig. 0.2. Scheme for synoptic diagrams of each ecozone

At the end of each regional chapter is a synoptic diagram showing the most important characteristics of the ecozones which enables comparisons between zones (Fig. 0.2). Figure 0.3 offers a direct comparison for some selected characteristics.

Ecozones / Characteristics	Polar subpolar (Ice deserts)	Tundra, Frost debris areas and areas without vegetation	Boreal	Temperate midlatitudes	Dry midlatitudes (Grass steppes, Desert, semi-desert)	Subtropics with winter rain	Subtropics with year-round rain	Dry tropics and subtropics (Thorn savanna, thorn steppes, Desert, semi-desert)	Tropics with summer rain	Tropics with year-round rain
Mean annual precipitation (P)	○	◕	◑	◕ ○	◑	◕	◕ ○	◕ ○	◕	●
Mean annual temperature	○	◕	◑	◑	◕	◕	◕	●	●
Mean annual potential evapotranspiration	○	◕	◑	●	●	◑	●	●	◕
Runoff (R)	◕	◑	◑	○	◕	◕	○	◑	●
Runoff coefficient (R/P)	●	◕	◑	○	◕	◕	○	◕	●
Annual solar radiation	○	◕	◑	◕	◕	◕	●	●	◕
Length of growing season	○ ◕	◑	●	◕ ○	●	●	◕ ○	●	●
Solar radiation during growing season	○	◕	◑	◕	◕	●	◕	●	●
Temperature during growing season	○	◕	◕	◑	◕	◕	◕ [2]	●	●
Biomass — Total	◕	◑	●	◕ ○	◕	●	◕ ○	◑	●
Biomass — Root-shoot ratio	●	◕	◕	◕ ●	◑	○	◑ ●	◑	○
Leaf area index	◕	●	◑	◑ ○	◕	◕	◕ ○	◑	●
Net primary production	○	◕	◑	◑ ○	◕	◕	◕ ○	●	●
Litter supply	●	●	◑	◕ ○	◑	◑	◕ ○	◕	◕
Dead organic soil material	●	◕	●	● ○	◑	◑	○	◕	◕
Decomposition rates	●	◕	◑	◕	◕	◑	◑	◕	○

Legend:
- ● Very high values
- ◕ High values
- ◑ Medium values
- ◔ Low values
- ○ Very low values or zero

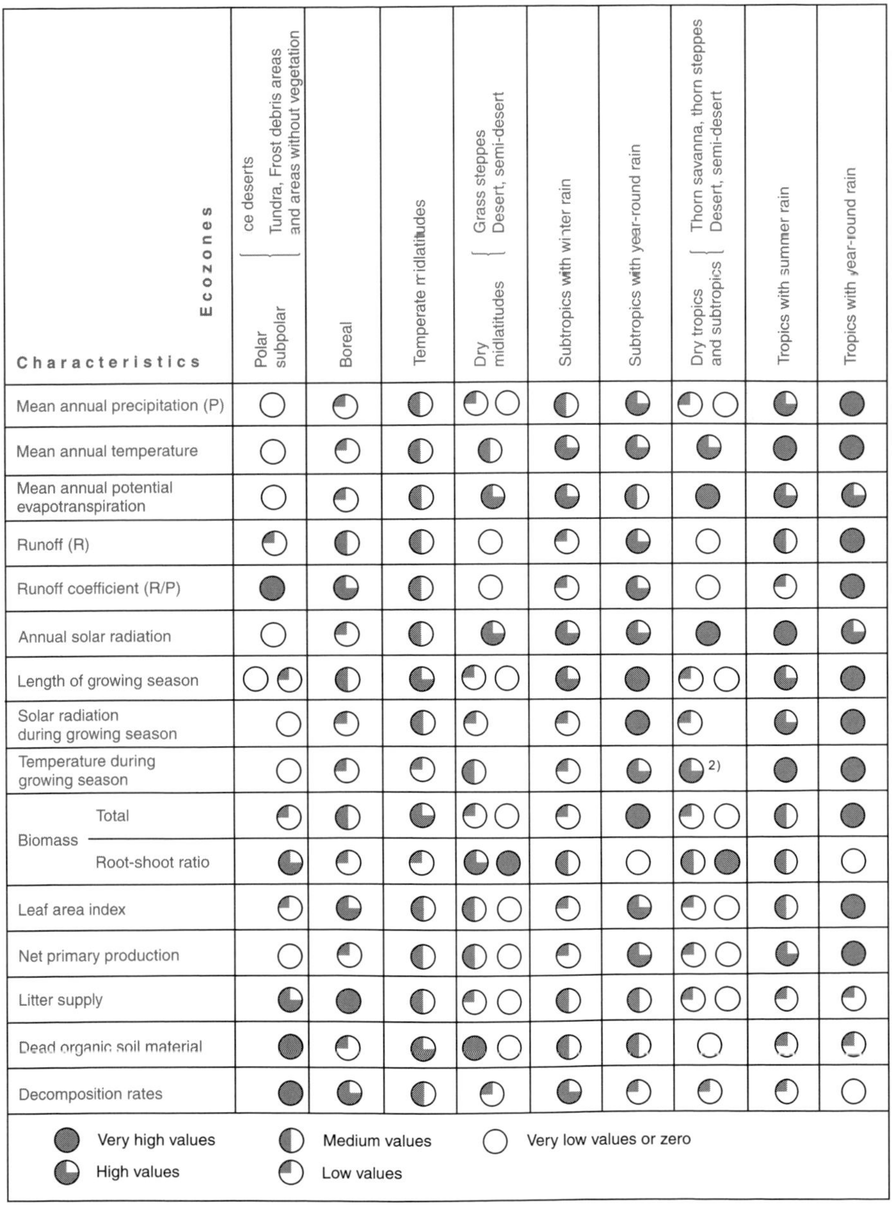

Fig. 0.3. Comparison of selected characteristics in the ecozones

The attempt to subdivide the world into a few ecological regions, i.e. eco-zones, which are highly uniform as regards their abiotic and biotic factors, the processes controlling their ecosystems, and their limits and prospects for human exploitation, is difficult for several reasons and has therefore been subject to criticism from some sides. The problems, which are difficult to solve and whose existence should not be denied, include the following:

a) *The wide range of small-scale variations in environmental conditions*, which exist everywhere on earth, can only be fitted into an ecozonal classification by applying many constraints and by accepting considerable blurring of the data as a result.

b) *A number of phenomena cannot be classified at all due to the lack of distinguishable environmental influences*; this group includes, for example, the land/sea interface, the overall relief of the earth, the distribution of rock types and mineral resources and many historically related phenomena (classification according to national boundaries, languages and so-called cultural groups). These characteristics and the effects they have, e.g. on the climate or the use of the land, disrupt the process of ecozonal classification, or they fail to fit into such a system at all.

c) *The remaining geographical elements, which are more or less dependent on the surrounding environment (and thus form part of the ecozonal web), seldom have distinct boundaries.* Typically, the changeover from one region to another is gradual and takes place over broad regions, in some cases along very different parameters (exceptions to this are the land/sea interfaces and boundaries formed by mountain ranges). The drawing of boundaries as distinct lines must, therefore, remain a questionable practice; all the more so when, as in the present case, they claim to be the boundaries delineating entire groups of characteristics.

d) *Many phenomena found on earth have developed over long periods of time.* For this reason their present form is the result of various environmental conditions to which the areas were subjected. Such regions can either not be classified at all or only with severe constraints.[1]

As a result of the problems just discussed:

1. ecozonal boundaries must to a certain extent be drawn arbitrarily (for example, along climatic thresholds) and are applicable only for a certain number of geographic characteristics; and

2. regardless of how the zonal boundaries are drawn, the variation in conditions in each of the zones must naturally remain large.

---

[1]Particularly in the case of landforms, the number of phenomena originating from the geological past is large. A morphogenetic explanation based on the processes predominating today is only possible on a very limited scale. Many soils also exhibit characteristics which have resulted over long periods (paleosoils); and most agricultural practices have strong historical roots. Much less prevalent are long-term characteristics in vegetation, since changes in environmental conditions cause rapid, sweeping changes in plant life (the postglacial changes of the forests in central Europe are a good example of this); and the climate is completely independent of elements from the past: the present climatic conditions are dictated solely by the global differentiation in energy supply from the sun, the earth's rotation and the existing tellurian and orographic conditions.

Nevertheless, an ecozonal classification of the world is possible and useful – that is the thesis of this book; however, it is subject to the following premises and concessions:

a) *Variations within the ecozones cannot be viewed as being inconsistent with their boundaries.* The decisive point is that at least some highly significant common elements do exist within each of the ecozones. Significance is measured in the case of characteristic factors according to their scale and functional dominance, and in the case of characteristic forms according to their distribution and conspicuousness. Significant factors and form characteristics in this sense are, for example, *inhibited decomposition of organic matter and thick raw humus layers on the ground* in the Boreal Zone or winter rain and *sclerophyllous (hard-leaved) vegetation* in the Subtropics with winter rain. To recognize common elements within zones, an adequate (global) yardstick is the most important prerequisite. When this tool is applied, many discrepancies disappear on their own accord. A figurative comparison may be able to clarify this point: when a small section of a small-scale world map is compared to a detailed, large-scale map of the same area, the former is found to be inexact and incomplete, and generalized to the point of being erroneous; but no one would argue that this map is of no value.

b) *Ecozones can only be characterized by the average conditions that predominate in them.* "Average" conditions are found in areas, which

– have no excessive runoff (erosion and denudation),

– exhibit neither considerable input (sedimentation) nor waterlogging,

– are at sea level or slightly above,

– have neither a distinctly continental nor a distinctly maritime climate.

Special cases in an orographic or edaphic sense can be included in the characterization if they are typical for a zone; this applies, for example, to Vertisols, saline soils or Histosols, which are typical final links in relief-related soil sequences (catenas) of the Tropic with summer rain, the arid lands and the Boreal Zone, respectively.

c) *The drawing of boundaries between the ecozones is of secondary importance.* The main emphasis must be placed on determining the (average conditions of the) core areas.

d) *All quantitative data can only be viewed as guidelines* (even when ranges are indicated, these do not reflect the actual extremes but rather the ends of the spectrum within which most data lie). The data are intended to make the global differences between the ecozones clearer and can serve as a measure for determining local deviations within each zone. The major aim of this book, by listing and characterizing the ecozones in the manner described, is to aid in establishing a kind of global classification system (orientation guide) which

– allows immediate listing of several important characteristics of any area on earth; and

– is suitable as a basis for further detailed investigation (starting with the question: in what way does a certain area in an ecozone differ from the overall characteristics of the ecozone in which it is found?).

Distribution

The size and spatial distribution of the ecozones is shown in Fig. 1.1 and Table 1.1 and in the regional chapters.

Table 1.1. Areal extent of the ecozones

Ecozones – Sub-ecozones	Area million km²	Landmass (%)
Polar subpolar	22.0	14.8
– Ice deserts	16.0	
– Tundra and areas of rock and frost debris	6.0	
Boreal	19.5	13.1
Temperate midlatitudes	14.5	9.7
Dry midlatitudes	16.5	11.1
– Grass steppes	12.0	
– Deserts and semi-deserts	4.5	
Subtropics with winter rain	2.5	1.7
Subtropics with year-round rain	6.0	4.0
Dry tropics and subtropics	31.0	20.8
– Deserts and semi-deserts	18.0	
– Winter-wet grass and shrub steppes (subtropics)	3.5	
– Summer-wet thorn savanna (tropics) and thorn steppes (subtropics)	9.5	
Tropics with summer rain	24.5	16.4
– Dry savanna	10.5	
– Moist savanna	14.0	
Tropics with year-round rain	12.5	8.4
Total area	149.0	100.0

The boundaries of the ecozones entered on the distribution maps adhere to the climatic zone structuring of the earth according to Troll and Paffen (1964), which do more justice to the earth zone's differentiation between vegetation and further natural characteristics than other effective climate classifications.

Nevertheless, their application for the drawing up of the ecozone remains makeshift, which in view of the fact that this book is primarily concerned

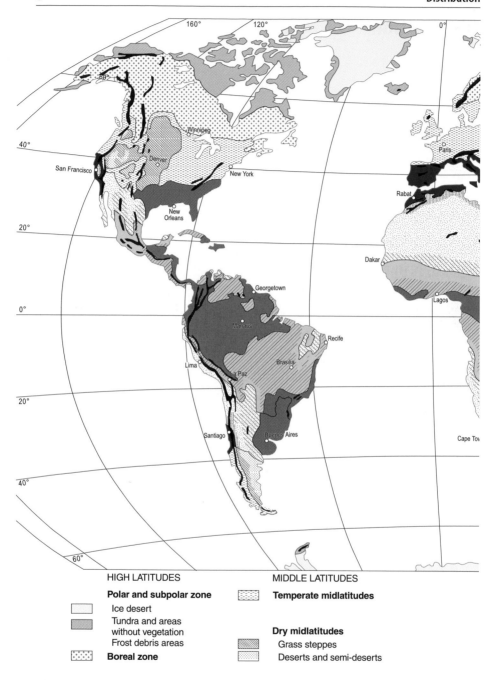

Fig. 1.1. Distribution of the ecozones of the world

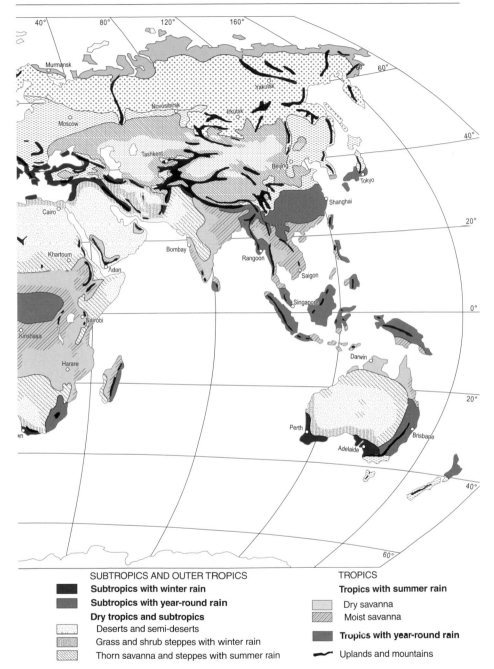

SUBTROPICS AND OUTER TROPICS

Subtropics with winter rain

Subtropics with year-round rain

Dry tropics and subtropics

Deserts and semi-deserts

Grass and shrub steppes with winter rain

Thorn savanna and steppes with summer rain

TROPICS

Tropics with summer rain

Dry savanna

Moist savanna

Tropics with year-round rain

Uplands and mountains

with the mean qualities of the ecozones renders the external boundary of less consequence and has to be accepted for the interim. Larger areas which are difficult to classify (e.g. thorn savannas) are referred to as "transitional areas" (Fig. 13.1 and Fig. 14.1).

CHAPTER 2

Climate

Climate provides the broad framework for the exogenic geomorphological processes, soil formation, plant growth and land use potential and is of fundamental importance in helping to define the character of an ecozone. Solar radiation and the length of the growing season are of particular importance for vegetation because radiation is the energy source for photosynthesis and determines the growing season length and therefore also the annual primary production of the vegetation.

The mean values of climatic data have only a limited usefulness for interpreting the effects of climate in an area. At least as important are the data on extreme events and their frequency, for example, data on precipitation intensities, the frequencies of extended periods of drought, strong winds, periods of intense cold and freeze thaw cycles. Also the micro climatic conditions within a small area and even within a plant cover may differ considerably from the climate for the ecozone as a whole.

2.1
Solar radiation

Global radiation is defined as the radiation with a shortwave range from about 290 to 3,000 nm that reaches the earth's surface as direct or diffused radiation (Fig. 2.1). The *photosynthetic active radiation (PAR)*, the radiation usable by plants, lies between 400 and 700 nm, the range for visible light. About half the energy from global radiation lies within these wave lengths.

The peak mean monthly radiation is similar in all ecozones (Fig. 2.2). Differences in annual growing season totals are caused by variations in the duration of the peak radiation and by variations in the time span within which plants can make use of the radiation energy they receive which, in turn, depends on their requirements for moisture and heat.

The *length of day* and its pattern of change throughout the year also affects the duration of the daily radiation. Ecozones in the middle and high latitudes are characterized by long hours of daylight and warmth in summer and by cold and darkness for long periods in winter. The low latitudes, by contrast, show little or no seasonal radiation and thermal differences. Energy transfers at the surface are shown in Box 1.

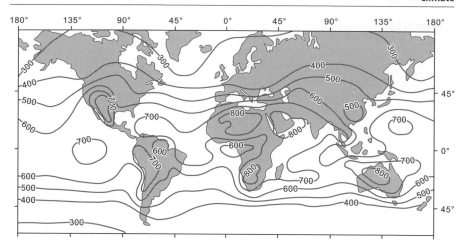

Fig. 2.1. Annual global radiation 10^8 kJ ha^{-1}. Source: De Jong 1973. Based on the uptake of this radiation it is possible to approximate primary production in the individual ecozones (Tab. 5.2)

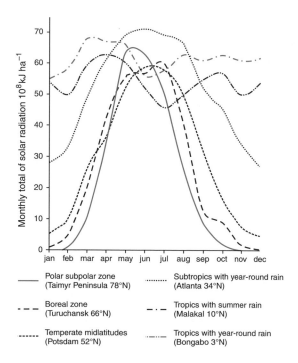

Fig. 2.2. Annual distribution of global radiation at weather stations in six ecozones

1

Simplified radiation (or energy) balance.[a]

$$(Q + q) \times (1 - \alpha) - A + G = LE + H$$

Q = direct solar radiation

q = diffuse radiation (downscatter)

α = reflected radiation as % of global radiation

} $Q + q$ = global radiation of which < 50% in photosynthetically active radiation (PAR)

α = albedo

} $(Q + q) \times (1 - \alpha)$ = absorbed short wave radiation (= net insolation)

$(Q + q) \times (1 - \alpha) - A + G$ = radiation balance (net radiation)

A = outgoing radiation (longwave)

G = counter radiation from within atmosphere (longwave)

} = temperature or heat radiation

} $A - G$ = effective longwave outgoing radiation (= net outgoing radiation)

LE = latent heat fluxes, LE is the transfer of heat energy connected with the state of water without changing the temperature, that is energy required for evaporation[b], melting[c], and/or release of heat by condensation and freezing (in immediate transition from solid to gaseous or solid state or vice versa: heat of sublimation)

H = sensible heat fluxes H is the temperature effective transfer of heat energy by molecular conduction at interfaces, includes heat transfers to and from the soil or snow cover to or from the atmosphere

[a] Lateral advective transport of energy, reflection of longwave counter radiation and photosynthetic energy fixed by plants are excluded. The latter ranges in terrestial areas from near zero in deserts to 0.8% in tropical rainforests of the global radiation.

[b] Energy required for evaporation or energy released by condensation of 1 g H_2O at 20 °C and normal pressure at sea level is 2.45 kJ

[c] Energy required to melt 1 g of ice or the energy released by freezing of 1 g H_2O is 0.17 kJ.

2.2
The growing season and conditions for plant growth

The *growing season* is defined as the annual total of months with a mean temperature of $t_{mon} \geq 5\,°C$ and precipitation (p) in millimeters of double the monthly mean value of the temperature in centigrade (C). This can be expressed as $p[mm] > 2t_{mon}[°C]$. Figure 2.1 shows the growing season in the ecozones derived from climatic diagrams in Walter and Lieth (1960–1967) and climatic tables in Müller (1996) and Landsberg (1970–86).

In the middle and high latitudes the plant growth is interrupted when the air temperature in winter falls low enough for one month or more to have a mean of $t_{mon} < 5\,°C$. In these areas a thermal seasonal climate dominates and the length of the cold period determines the length of the growing season. In tropical and subtropical ecozones temperatures do not fall low enough to limit growth, the annual range is in fact smaller than the diurnal range, but variability in the supply of moisture causes periods of drought which interrupt growth and affect the length of the growing season.

Table 2.1. Thermal and moisture conditions for growth in the ecozones

Ecozones	Growing season Months with $p(mm) > 2t_{mon}(°C)$ and $t_{mon} \geq 5\,°C$	Months with $t_{mon} \geq 10\,°C$	$t_{mon} \geq 18\,°C$	Annual precipitation (mm)
Polar subpolar	0–3 (4)	0 (1)	–	< 250
Boreal	4–5 (3–6)	2–3 (1–4)	0 (1)	250–500
Temperate midlatitudes	6–12 (5)	5–7 (4)	1–3 0–5	500–1000
Dry midlatitudes	0–4 (5)	5–7	≤ 4 (5)	< 400 < 200 (250) (summer)
Subtropics with winter rain	6–9 (5–10)	8–12	4–6	500–1000
Subtropics with year-round rain	12	8–12	4–7 (up to 12)	1000–1500
Dry tropics and subtropics	0–4 (5)	12 (9–11)	5–12	< 300 polewards < 500 equatorwards
Tropics with summer rain	6–9 (5)	–	12	500–1500
Tropics with year-round rain	12	–	12	2000–4000

Values in parenthesis are regional exceptions due to continental, maritime or latitudinal differences

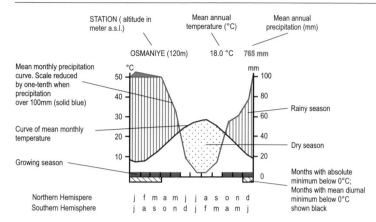

Fig. 2.3. Climatic diagram for a weather station in Turkey in the Subtropics with winter rain. The mean monthly precipitation in mm and the mean monthly temperature in °C are shown in a relationship of 1:2 i.e., 20 °C is equated with 40 mm precipitation. When the precipitation curve is above the temperature curve, the months are moist, when below, arid.

In addition to the length of the growing season, the available radiation energy and the level of the air temperature are significant for plant growth. The mean monthly temperature in the tropics is over 18 °C in all months and in the subtropics in at least four months. On the polar boundary of the Boreal zone, a mean of 10 °C is reached in only one month (Tab. 2.1).

CHAPTER 3

Relief and drainage

3.1
Geomorphological processes

The chapters on relief primarily show which exogenic geomorphological processes are characteristic (currently effective) and which land surface formations can be attributed to them. If need supplementary information on such landforms will be presented which result from tectonic movements (endogenic processes such as earth crust movements and vulcanology) or which are the product of earlier morphoclimatic conditions. There are numerous such landforms in the earth's relief which without exception dominate the landscape, but which defy all ecozonal classification.

The morphodynamics differ in the ecozones more or less directly in relation to the type, frequency, duration and intensity of the processes

- weathering due to splitting by frost, temperature or salt, or due to chemical processes such as hydration, solution, oxidation or hydrolysis, and

- erosion and deposition caused by water, ice, wind or solely by gravity (fluvial erosion, marine abrasion, glacial erosion, deflation and denudation etc.)

The weathering products and the erosional and deposition forms that result depend largely on the precipitation, temperature and wind regimes and on the bedrock at the land surface. Figure 3.1 shows in a profile from the poles to the equator, the depth, structure and chemical characteristics of the weathering cover in relation to the climate. The fact that exogenic processes are largely a function of climate means that a subdivision of the earth into regions with similar geomorphological processes is similar to the global climatic patterns and to the distribution of the ecozones (Stoddart 1969, Büdel 1981, Tricart and Cailleux 1972, Hagedorn and Poser 1974).

3.2
Drainage and water balance

The most effective geomorphological processes in all ecozones except sand deserts and the inland ice, are those related to water channelled in streams or to overland flow on slopes. The volume and velocity of the flowing water

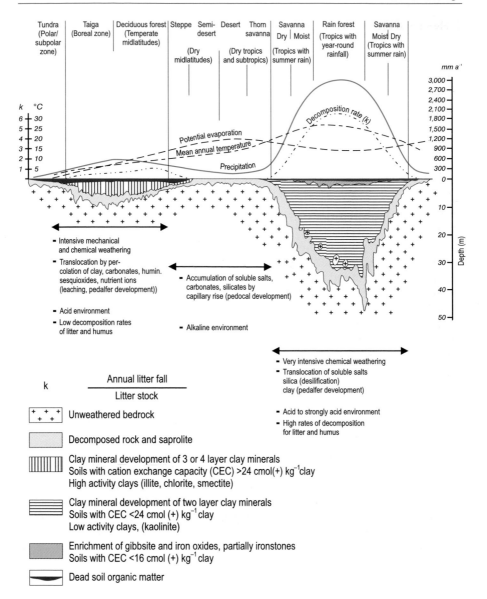

Fig. 3.1. Variations in the weathered cover (regolith and saprolite) in climatic zones

Table 3.1. Mean annual runoff and runoff ratios in the ecozones

Ecozones	Formation type	Annual runoff (mm)	Runoff ratio (%)
Polar subpolar	Tundra	120	55
Boreal	Taiga	200	50
Temperate midlatitudes	Deciduous forest	350	47
Dry midlatitudes	Tall grass steppes	200	40
	Short grass steppes	60	12
	Desert and semi-desert	< 10	< 3
Subtropics with winter rain	Sclerophyllous vegetation	300	50
Subtropics with year-round rain	Rainforest	650	43
Dry tropics and subtropics	Thorn savanna	50	8
	Desert and semi-desert	< 5	< 3
Tropics with summer rain	Dry savanna	250	33
	Moist savanna	450	45
Tropics with year-round rainfall	Rainforest	1200	52
World mean		310	41

Runoff ratio = % of annual precipitation in the annual runoff

Source: UNESCO 1978, Henning 1989 et al.

determines the extent and intensity of the processes. The volume (discharge) is a function of the area of the drainage basin and its water balance and is expressed as m^3 [or l] sec^{-1} or $mm\,a^{-1}$.

Rain and snow supply the drainage basin (Fig. 4.3). How much reaches the streams is determined from the long term means of the difference between precipitation and evaporation in a drainage basin or larger area. Long term storage at depth is taken into account. Estimates of storage for short periods measure changes in the soil water, groundwater and surface water.

Table 3.1 shows approximate values for the total discharge in each ecozone and the ratio to total precipitation (drainage quotient) in the ecozone. Table 2.1 shows typical precipitation.

CHAPTER 4

Soils

Compared to climate, soils are of less importance in defining natural regions but are sometimes dominant at a spatially smaller scale.

4.1
Soil fertility

Soil fertility is related to the supply of basic nutrients, how they are bound chemically and their quantity, and to the climate dependent heat, air and water budgets of the soil which are significant, and sometimes the dominant factor in determining the productivity of the soil.

The quantity of *available nutrients* in the soil depends on the size of the exchangeable fraction which, in turn depends on the *exchange capacity* of the soil and the balance of nutrients added or lost to the soil (Fig. 4.1). Where there is plant cover, the *nutrient balance* is largely determined within the ecosystem in which mineral nutrients become available following decomposition in the soil, and are withdrawn into the biomass by plant growth. There are also imports and exports external to the system including the long term release of silicates from the bedrock.

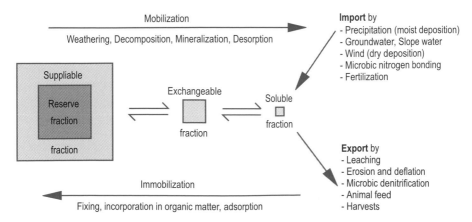

Fig. 4.1. Nutrient fractions and relevant processes. Source: after Schroeder and Blum 1992

The exchange capacity, i.e. the ability to adsorb nutrient ions due to excess negative or positive charges of some of the soil components, is coupled to the amount and types of clay minerals and the *soil organic matter* (SOM). The *cation exchange capacity* (CEC) which is a measure for the exchange of the major nutrient elements Ca^{++}, K^+, Mg^{++} reaches its highest value in *mull* (a type of humus) with $200-500$ cmol(+) kg^{-1}. Among the clay minerals the three layered smectite and vermiculite have the highest values with > 100 cmol(+) kg^{-1}, followed by the three layered clay mineral illite with 20 to 50 cmol(+) kg^{-1} and

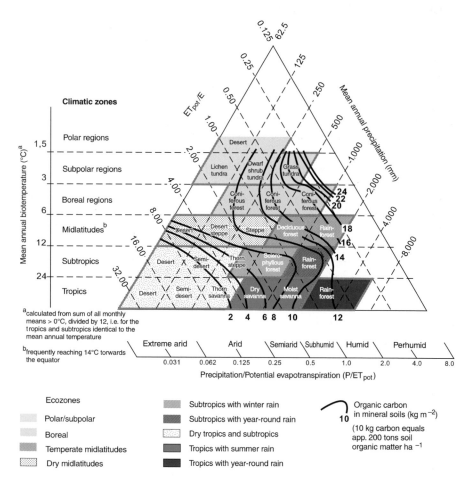

Fig. 4.2. The carbon supplies in mineral soils (soils without litter layer and components of more than 2 mm in diameter) in relation to climate and vegetation (Post et al. 1982). The carbon content was based on 2700 soil profiles, subdivided on the basis of plant formations (*life zones* after Holdridge 1947), from which the mean carbon was derived and plotted. The increase in carbon shown in the figure from left to right indicates an increase in production with increasing precipitation. The increase in carbon from the bottom of the figure towards the top indicates decreasing rates of decomposition with decreasing temperature.

the two layer kaolinite with 5 to 15 $cmol(+) kg^{-1}$. Sesquioxides, i.e. oxides and hydroxides of iron and aluminum such as goethite, hematite and gibbsite have a very low, largely anion, exchange capacity (AEC), for example, NO_3^- SO_4^{--} and PO_4^{---}.

For this reason, Vertisols, in which smectite dominates in their clay fractions, have a high cation exchange capacity. Other tropical and subtropical soils, such as Lixisols, Acrisols and Nitisols, in which kaolinite dominates, have low cation exchange capacities of < 24 $cmol(+) kg^{-1}$.

Extremely low cation exchange capacity is found in soils in which in addition to kaolinite there is also a high proportion of sesquioxides , as found for example in some Ferralsols (where as little as < 1.5 $cmol(+) kg^{-1}$ clay have been measured). In the case of Middle European soils the cation exchange capacity ranges between 40 and 60 $cmol(+) kg^{-1}$ clay.

Soil fertility benefits when nutrient ions have a high share of the cation exchange capacity and therefore a high *base saturation* and also if most animal and plant detritus is mineralized within a few years so that the nutrient elements are again available for plant growth. In addition, fertility improves if a mull develops in the form of a well decomposed fine humus in which there is bonding with clay minerals.

Fertility depends, therefore, on how much dead organic matter reaches the soil, which is dependent on net primary production, how easily litter is decomposed and the conditions that exist for decomposition to take place, particularly the environment for *soil organisms* (Chap. 5.6). Soil organisms can be restricted by drought, cold, soil acidity or lack of oxygen due to saturation of the soil. Because of the interaction of all these factors, the humus content, the type of humus and the dynamics of nutrient exchange varies considerably from one ecozone to another (Fig. 4.2).

4.2
Soil water budget

The *soil water budget* is the state of the soil water according to distribution, storage and bonding as well as its changes in time. Water is largely supplied to the soil by *infiltration*. In exceptional cases *groundwater* may be the main source of supply. The total infiltration in an area may be smaller or larger than the total precipitation that falls on the soil surface. On plant covered surfaces, some part of the precipitation does not percolate downwards but is lost to evaporation. In dry areas with little or no vegetation cover the soil water supply comes largely from surface runoff, or even condensation, rather than from precipitation. Figure 4.3 shows the distribution of precipitation falling on a surface.

Water infiltrating either remains in the soil as soil moisture or percolates down into the groundwater. The *field moisture capacity* is the maximum amount of water held in the soil after excess moisture has flowed away. It is sometimes defined as including water percolating slowly down and held in

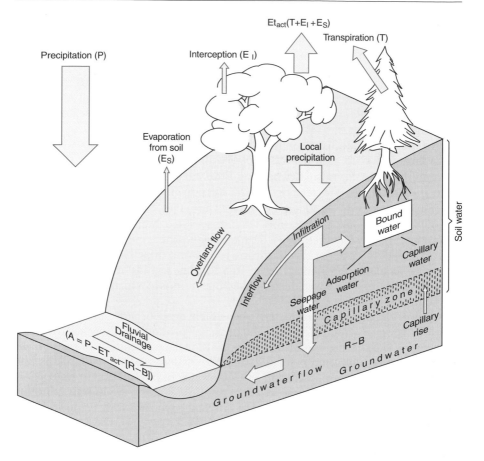

Fig. 4.3. Distribution of precipitation and soil water. The arrow width is in proportion to the annual flows in the Temperate midlatitudes. Totals and proportions vary greatly in other ecozones. R-B is the change in total groundwater, R = bound water, groundwater or snow; B = consumption

medium sized pore spaces of 10 to 50 μm, water which is therefore periodically usable (Fig. 4.4).

The field capacity is reached only temporarily and is confined to individual soil horizons. In the Temperate midlatitudes, for example, the summer growing season may use all the water reserves stored in the soil in winter when field capacity was reached (Fig. 9.5). In the Subtropics with winter rain, and the Subtropics with year-round rain the annual reversal from reserve to consumption occurs with the change from rainy to dry seasons. In dry areas the field capacity is rarely reached, particularly in the subsoil, with exceptions perhaps in some of the steppes in the midlatitudes. Only in the subtropical and tropical rainforests do the soils hold more or less the maximum amount of water for the entire year. After the spring melt and subsequent thaw of the surface soil

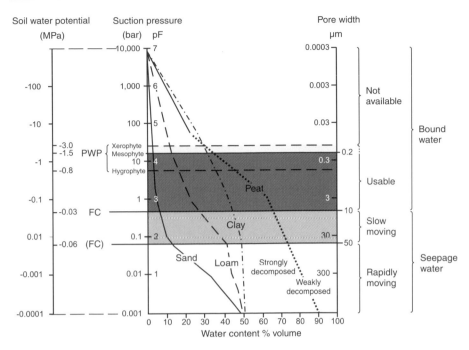

Fig. 4.4. Suction pressure of different soil types (PWP = Permanent wilting point, FC = Field capacity, pF = Log of suction pressure (Water tension in cm of water column)

layers in the Polar subpolar and parts of the Boreal zones the soil becomes saturated, especially where permafrost prevents drainage downwards.

Plants can use the remaining soil water only when they have built up a high *root suction potential* or the osmotic potential in their roots sinks further and they can continue to draw water up into their hair roots. When there is an equilibrium, the *permanent wilting point* is reached. For most plants the wilting point is at a soil water tension of about 15 bars, or a matrix potential of −1.5 MPa. Plants in dry areas have lower root potentials, to about −6 MPa.

Field capacity and permanent wilting point form the absolute boundaries for the usable field capacity. Figure 4.4 and Table 4.1 show that both the field capacity and the permanent wilting point, in relation to soil type, defined on the basis of the grain size ratio in the fraction, are reached at quite different water contents and the maximum possible content of usable water varies greatly. The highest and most advantageous are contents of up to 20 vol.% in loam and silt, moderately high and favorable are contents of about 15 vol% in clays. Lowest are sands with about 7 vol.%

The actual amount of water available to plants depends also on the size of the *rhizosphere* (root area) and the *root penetration* intensity. The maximum amount that would be available, the water use capacity in the root area, is derived from the difference in water content between the field capacity and the permanent wilting point multiplied by the depth of the root system and taking into account the root penetration intensity (Table 4.1).

Table 4.1. Available water in root area of plants in selected soils

Soil types	Mean water content FC	PWP	Usable water content FC – PWP	Depth of root area	Usable water capacity in root area
	volume (%)		volume (%)	(dm)	(mm)
Cambisol (sandy)	10	3	7	10.0	70
Chernozem (silty)	30	10	20	15.0	300
Luvisol (loamy)	35	15	20	7.5	150
Clay soil (Vertisol)	45	30	15	5.0	75

FC = Field capacity, PWP = Permanent wilting point
Water content = mean value of all horizons in root area
1 mm soil water equals 1 volume% water in a 1 decimeter thick soil layer

In a sandy soil layer only 70 mm precipitation can be stored in the root area if the PWP is taken as a base value; using the field capacity, 100 mm are required. Larger quantities than these do not benefit the plant growth.
The water in this type of soil is used up rapidly so that precipitation events must occur at relatively short intervals. In a silty soil, 300 mm of precipitation can be stored to supply the plants. In this case intervals between precipitation events can be longer.

Source: Schroeder and Blum 1992

4.3
Soil units and soil zones

The *FAO classification of soils* is used in this book. The classification is the basis for the World Soil Map at a scale of 1:5 million with 18 sheets (FAO-UNESCO 1974–81). The map was produced from regional soil maps, which are also the basis for a European soil map at 1:1 million.

Box 2 describes briefly the major soil groupings and some soil units in the FAO Classification together with the comparable soil types of the *U.S. Soil Taxonomy*. Exact comparisons are usually not possible because the defining criteria are different.

The names of the soil units are related to the characteristics of formative elements of the soils, for example,

- Albic – bleached
- Calcic – secondary calcium enrichment
- Chromic – strongly colored
- Dystric – infertile, low base saturation
- Eutric – fertile, high base saturation
- Ferralic – high proportion of sesquioxides (Fe and Al oxides/hydroxides)
- Fibric – rich in partially decomposed organic material (peat)
- Gelic – permafrost in subsoil
- Haplic – normal sequence of horizons

Table 4.2. Soil zones and ecozones

Soil zones	Ecozones/Subdivisions
Gelic Regosol – Gelic Gleysol	Tundra and areas of rock and debris
Podzol – Cambisol – Histosols	Boreal
Haplic Luvisol	Temperate midlatitudes
Kastanozem – Haplic Phaeozem Chernozem	Grass steppes (moist)
Xerosol	Grass steppes (dry)
	Thorn savanna, thorn steppes
Yermosol	Desert and semi-desert in midlatitudes, subtropics and tropics
Chromic Luvisol – Calcisol	Subtropics with winter rain
Acrisol – Luvisol – Nitisol	Tropics with summer rain
Acrisol	Subtropics with year-round rain
	Tropics with year-round rain in SE Asia
	Moist savanna in South America
Ferrasol	Tropics with year-round rain, excluding SE Asia and Central America

Xerosols and Yermosols have been replaced in the FAO classification of 1988 by other soils, some newly defined (see Figure 4.5). Additional changes have been made in 1998. The Gelic-Regosol – Gelic Gleysol zone is renamed Cryosol zone, the Podzol-Cambisol-Histosol zone as Podzol-Umbrisol-Histosol zone and the typical soil for Xerosols are Durisols.

- Luvic – clay accumulation in subsoil (argic B horizon or Bt horizon)
- Plinthic – Fe rich clays (rusty flecks), petroplinthic if hardening irreversable
- Umbric – rich in humus, base saturation < 50%.

The legend of the FAO-UNESCO map was revised in 1988 and 1994 (FAO 1988 and 1994, see also Driessen and Dudal 1991). Another revision was published in the World Reference Base for Soil Resources in 1998 (ISSS, ISRIC, FAO, Bridges et al., Deckers et al.). The important changes have been incorporated. Where necessary, the older names for the soils are retained for clarity.

The World Reference Base for Soil Resources Map has introduced Cryosols, Durisols and Umbrisols as major soil groups and renamed Podzoluvisols, as Albeluvisols and Greyzems as Phaeozems. Cryosols are soils that have permafrost at a depth of less than 1 m and evidence of cryoturbation in their structure. The Gelic Gleysols, Gelic Regosols, Gelic Leptosols etc. terms used in this book, are all Cryosols.

Durisols are characterized by duricrusts or nodules of silcrete. They appear as Xerosols on the FAO-UNESCO maps. Umbrisols are humus rich soils with a base saturation of less than 50%, in contrast to the humus rich steppe soils which have a higher base saturation. Umbrisols occur mainly in the Boreal zone.

Soil Classification   2

FAO Classification[a/] Major soil groupings	Characteristics	US Soil Taxonomy and Classic soil groups
1 Fluvisols	Young soils in fluvial deposits (flood plains) undeveloped profiles	Entisols: Fluvents Alluvial soils
Gleysols	Soils influenced by high water table (hydromorphic soils), not on flood plains.	Entisols: Aquents; Inceptisols: Aquepts Mollisols: Aquolls; Meadow soils
– Gelic		Pergelic Cryaquepts, Tundra gleysols
Regosols	Soils without horizon development on clayey or loamy sediments, not on flood plains	Entisols: Orthents
Leptosols[b/]	Weakly developed shallow soils on bedrock or rocky debris (include former Rankers, Rendzinas, Lithosols and stony Regosols).	Entisols: Lithic subgroups; Rendolls
Lithosols[c/]	Shallow soils, < 10 cm, without horizon, on bedrock	Entisols: Orthents
Rankers[c/]	Shallow AC soils with umbric A horizon on rocks free or poor in carbonate	Inceptisols: Lithic Haplumbrepts
Rendzinas[c/]	Shallow AC soils with mollic A horizon on rocks rich in carbonates.	Mollisols; Rendolls
2 Arenosols	Sandy soils except on flood plains, with weakly developed profile. Ochric A horizon (poor in humus)	Entisols: Psamments, Psammaquents; Red and yellow sands.
Andosols	Young, loosely structured dark humus-rich soils on volcanic ash	Andisols; Black volcanic soils, Humic allophane soils
Vertisols	Dark clay soils rich in smectite with pronounced swelling and shrinking (gilgai, cracking, self-mulching effect).	Vertisols, Tirs, Black cotton soils, Cracking clay soils, Grumusols

				2

Soil Classification (continued)

FAO Classification[a/] Major soil groupings	Characteristics	US Soil Taxonomy and Classic soil groups
3 Cambisols	Soils with loamy subsoil and in situ oxidation of iron. Brown color. Weathered cambic B horizon	Inceptisols: Ochrepts, Umbrepts; Brown earths.
– Calcaric		Brown earths
– Chromic		Rhodic Xerochrepts
– Dystric		Dystrochrepts, Acid brown forest soils
– Eutric		Eutrochrepts, Orthic brown forest soils
4 Calcisols[b/]	Weakly developed soils with ochric A horizon, usually in dry areas with secondary calcerous enrichment (secondary calcium)	Aridisols: Calciorthids; Calcareous soils, Calcic and petrocalic subgroups of desert soils
Gypsisols[b/]	Weakly developed soils with ochric A horizon, usually in dry areas with secondary gypsum enrichment	Aridisols: Gypsiorthids; Soils with gypsum
Solonetz	B horizon with clay enrichment and high degree of natrium (sodium) saturation (natric B)	Aridisols: Natrargids; Natric Alkali soils, Sodic soils
Solonchaks	Soils with high content of soluble salts	Aridisols: Salorthids; Saline soils
Xerosols[d/]	Humus poor soils with ochric A horizon in semi-desert	Aridisols, Semi-desert soils, Burozems
Yermosols[d/]	Soils extremely poor in humus with ochric A horizon in deserts	Aridisols, Desert soils.
5 Kastanozems	Steppe soils with chestnut colored humus rich mull A horizon on calcareous or gypsum containing subsoil	Mollisols: Ustolls and Aridic Borolls; Chestnut soils in the dry steppe, Brown and dark soils
Chernozems	Steppe soils with thick dark mull A horizon above calcareous subsoil, intensive bioturbation	Mollisols: Borolls; Black earths of temperate steppes, Black soils

Soil Classification (continued)		2

FAO Classification[a/] Major soil groupings	Characteristics	US Soil Taxonomy and Classic soil groups
5 – Haplic		Haploborolls, Typic Chernozems
– Luvic		Agriborolls, Podzolized Chernozems
Phaeozems	Degraded steppe soils; dark humus rich mull A horizon above decalcified subsoil	Mollisols: Udolls. Dark grey soils. Brunizems.
Greyzems[e/]	Grey soils with mull A horizon and bleached aggregate surfaces in clay enriched B horizon (argic B horizon)	Mollisols: Agriborolls, Aquolls, Grey forest soils
6 Luvisols	Leached soils with high CEC (> 24 cmol(+)kg^{-1} clay) and base saturation > 50 %	Alfisols: Udalfs, Boralfs; Grey brown podzolic soils
– Chromic		Rhodoxeralfs, Haploxeralfs
– Haplic	(formerly Orthic)	Podzolized brown forest soils
Planosols	Poorly drained soils with wet bleached E horizon and largely impermeable subsoil marbled with iron flecks	Albaqualfs, Albaquults, Argialbolls
Podzoluvisols[f/]	Leached soils with bleached E horizon tonguing deep into the illuvial B horizon	Alfisols: Glossic great groups
Podzols	Leached albic E horizon and black or rust brown spodic B horizon with accumulation of humin material, Al and Fe	Spodosols
7 Lixisols[b/]	Leached soils with low CEC (< 24 cmol(+)kg^{-1} clay) and high base saturation > 50 %. CEC distinguishes them from Luvisols, the base saturation from Acrisols	Alfisols: Ustalfs, Xeralfs; Ferruginous soils, Red-yellow podzolic soils
Acrisols	Strongly weathered, acid, kaolin soils with argic B horizon, low CEC (< 24 cmol(+)kg^{-1} clay), base saturation < 50 %	Utisols, Red-yellow podzolic soils, Leached ferralitic soils

Soil Classification (continued)		2

FAO Classification[a/] Major soil groupings	Characteristics	US Soil Taxonomy and Classic soil groups
7 Alisols[b/]	Similar to Acrisols but with higher CEC (≥ 24 cmol(+)kg^{-1} clay), high Al saturation and therefore low base saturation	Aquults, Humults, Udults
Nitisols	Soils with nitic B horizon with low CEC (< 24 cmol(+)kg^{-1} clay), stable polyhedron structure with glossy aggregate surfaces (formerly Nitosols)	Alfisols: Udalfs; Ultisols: Udults, Ustults; Krasnozems
8 Ferralsols	Deeply weathered ferralitic soils (rich in Al and Fe oxides) with stable structured ferralic B horizon, very low CEC (≤ 16 cmol(+)kg^{-1} clay), Pseudosand	Oxisols, Weathered ferralitic soils, Latosols, Red earths, Lateritic soils
Plinthosols[b/]	Iron rich soils, usually with rusty flecks which harden when dry out (Laterite)	Oxisols: Plinthaquox; Latosols, Groundwater laterite soils
Histosols	Organic soils with organic horizon at least 40 cm thick	Histosols, Bog soils, Peat soils, Organic hydromorphic soils
Anthrosols[b/]	Soils altered by man	Anthropogenic soils Agrozems

a/ All 28 major soil groupings of the Revised Legend (Rome 1988) and some of the soil units. Also included are major soil groupings which appear in the FAO-UNESCO soil map but which have since been excluded (Lithosols, Rankers, Rendzinas, Xerosols Yermosols, Greyzems). With the introduction of Cyrosols, Umbrisols and Durisols and the exclusion of Greyzems, the reference soil groups now total 30.

1–8 are related to the soil genetic and soil ecological similarities. The orders of the US Soil Taxonomy are shown in parenthesis.

1. Young soils, present in all climates (Entisols)
2. Bedrock related soils (Entisols, Vertisols)
3. Weakly developed soils with B_w horizon in all climatic zones (Inceptisols)
4. Soils with salt enrichment, dry climate (Aridisols)
5. Soils with deep mull A horizon; steppes (Mollisols).
6. Soils with clay or humus enrichment in B horizon, medium to high base saturation, outside tropics (Alfisols, Spodosols).
7. Deeply weathered soils with clay enrichment in the B horizon, most have low CEC and base saturation; tropics and subtropics (Utisols, Alfisols)
8. Deeply developed Si poor soils without clay accumulation, rich in sesquioxides, humid tropics (Oxisols).

b/ Added since 1988 (Revised Legend 1988).
c/ Belong to Leptosols since 1988. Lithosols = Lithic Leptosols; Rankers = Umbric Leptosols; Rendzinas = Rendzic Leptosols.
d/ Replaced in 1988 by Calcisols, Gypsisols, Arenosols, Regosols etc.
e/ Replaced by Phaeozems in 1998
f/ Renamed Albeluvisols in 1998

Source: FAO-UNESCO (1974–1981) Soil Map of the World, Vol. I–X and 18 maps, 1:5 million, UNESCO, Paris.
FAO (1988, 2nd Edition 1990). Revised Legend of the FAO-UNESCO Soil Map of the World. World Soil Resources Rep. 60, Rome, 119pp.
FAO (1994): Soil Map of the World – revised legend with corrections. ISRIC Technical Paper 20, Wageningen, 140pp.
USDA Soil Survey Staff (1999): Keys to soil taxonomy, Pocahontas Press. Blacksburg, VA, 3rd edition, 600pp.

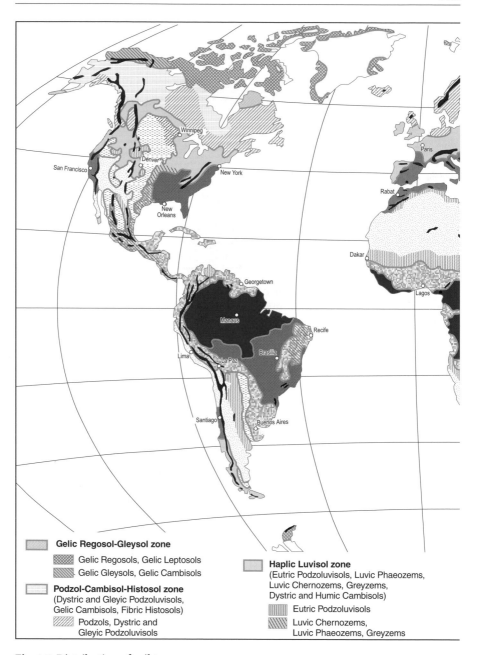

Fig. 4.5. Distribution of soil types

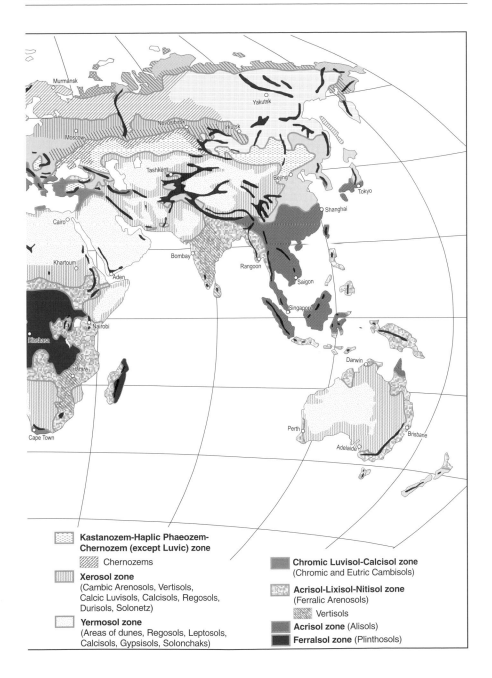

Kastanozem-Haplic Phaeozem-Chernozem (except Luvic) zone

Chernozems

Xerosol zone
(Cambic Arenosols, Vertisols, Calcic Luvisols, Calcisols, Regosols, Durisols, Solonetz)

Yermosol zone
(Areas of dunes, Regosols, Leptosols, Calcisols, Gypsisols, Solonchaks)

Chromic Luvisol-Calcisol zone
(Chromic and Eutric Cambisols)

Acrisol-Lixisol-Nitisol zone
(Ferralic Arenosols)

Vertisols

Acrisol zone (Alisols)

Ferralsol zone (Plinthosols)

CHAPTER 5

Vegetation and animals

Despite intervention by man, the vegetation cover and to a lesser extent the animal world reflect more than any other physical characteristics the ecozonal differentiation. For this reason, the vegetation and fauna are discussed in greater detail than other physical factors in the chapters on the individual ecozones and the components of the vegetation and fauna are described in order to show what the natural environment for plants, animals and man is like and also to indicate the risks and potentials for future development.

5.1
Structural characteristics of the vegetation

The structural characteristics of the vegetation cover include height and density of the cover, expressed by size parameters such as the *leaf area index*, the number of *trunks* and the *basel surface*. Other structural elements include measurements related to the *root penetration* depth and intensity, the species

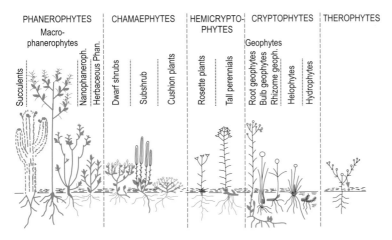

Fig. 5.1. Growth forms. Those parts of the plants drawn in blue survive dry seasons or cold seasons, the remainder die off at the beginning of the dry or cold season. Source: Raunkiaer (from Schubert 1991)

Table 5.1. Plant formations in the ecozones

Plant formations (Climax formations)	Ecozones
Polar desert High arctic tundra Low arctic tundra	Polar subpolar
Forest tundra Lichen forest Closed boreal coniferous forest – Evergreen boreal coniferous forest (dark taiga) – Deciduous boreal coniferous forest (light taiga)	Boreal
Deciduous and mixed forest Temperate rainforest – Evergreen deciduous and mixed forest – Temperate coniferous forest	Temperate midlatitudes
Woodland steppe Tall grass steppe Mixed grass steppe Short grass steppe Desert steppe Temperate desert	Dry midlatitudes
Sclerophyllous forest and shrub formations	Subtropics with winter rain
Subtropical rainforest Laurel forest	Subtropics with year-round rain
Winter wet grass and shrub steppe Summer wet thorn steppe and thorn savanna Tropical and subtropical desert and semi-desert	Dry tropics and subtropics
Short grass savanna (dry savanna) and dry forest High grass savanna (moist savanna) and moist forest	Tropics with summer rain
Tropical rainforest	Tropics with year-round rain

make up in a community, their growth patterns and their spatial distribution. The temporal dimension includes seasonal change, long term cycles in the stand including, aging and *regeneration*, and *plant succession*.

The convergence between the biozonal and ecozonal patterns is based on the convergent development of different plant species which takes place world wide as they adapt to the conditions at a particular location. A small number, compared to the total number of species in the world, of *life-forms* or growth forms have evolved to be similar in appearance and function in the ecosystem, despite differences in their taxonomies. Cactus in the Americas and succulent euphorbias in Eurasia are an example.

Plant cover can be described either in terms of a plant community, which is the species mix within a community, or in terms of a life-form spectrum which measures the relative share of the individual life-forms and their cover

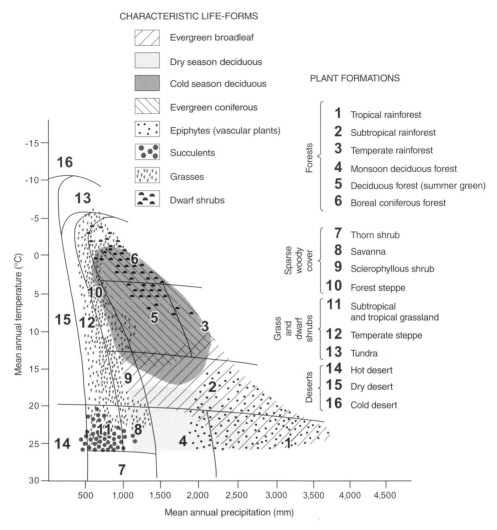

Fig. 5.2. Distribution of life-forms and plant formations in relation to annual temperature and annual precipitation. Source: from Sitte et al. 1998

within a community. The latter are *plant formations* based on physiognomic-ecological units of vegetation which, to a considerable extent, express the abiotic environmental conditions. Figure 4.2 shows the relationship to climate of the ecozones.

The zonal formations are the largest units of formation types and include boreal coniferous forests, deciduous and mixed forests and evergreen sclerophyllous forests. Their natural development is an expression of the global differentation of climates. They form the climax vegetation. Ecozones are, or in some cases were, represented by one or more *climax formations* (Table 5.1).

A biome includes both the plant formations and the animals within an ecosystem, a zonobiome is the combination of plant formations and animals within an ecozone. Transition zones between two communities are termed ecotones, in its largest extent a zonoecotone.

The growth forms of vegetation have been classified in various ways. Raunkiaer's classification is shown in Fig. 5.1. Figure 5.2 shows which growth forms are characteristic for individual plant formations and the distribution of different plant formations in relation to mean annual temperature and mean annual precipitation.

5.2
Ecosystem model of ecozone

A natural of near natural *ecosystem* is composed of a *biotic* community of plants and animals, or biocoenosis, and its *abiotic* environment, the biotope. A wide variety of functional and structural relationships exist between biotic communities and their biotopes. Conditions, apparently undisturbed from outside, develop a fairly stable structure in which self-regulation and self-regeneration take place and turnovers of materials and energy reach a form of equilibrium in which reserves of organic matter and minerals are constant.

Because ecosystems are dynamic systems in which the individual complexes such as vegetation, fauna or microclimate change continually, there are no real dynamic equilibria and constancy of components. Instead they are derived as mean states, either temporal, such as *aging* or *regenerating cycles* which are present everywhere in an ecosystem, or spatial states which occur all over an ecosystem in mosaic-like pattern of complexes of different ages.

Aging and regeneration cycles are particularly noticable in forests especially in the Boreal zone, the Temperate midlatitudes, the Subtropics with winter rain, and the Subtropical and Tropical zones with year-round rain. In these zones, a mature or optimal phase during which the highest values for primary production are reached, is followed by an aging or decomposition phase during which gaps are left by the dying trees. The vegetation that regenerates in these gaps are the *pioneer plants* that are characteristic for this rejuvenation phase.

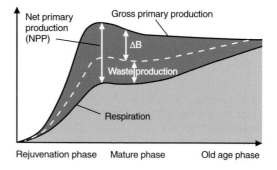

Fig. 5.3. Age-dependent changes in primary production, increment (ΔB), waste production and respiration in a forest formation. Source Kira and Shidei 1967

In Fig. 5.3, the highest net primary production is reached in the transition phase before maturity, when the total increment is also largest. Respiration then also increases relatively rapidly as the relationship between productive leaves and unproductive stems, branches and roots is less favorable and the net primary production declines again. In deciduous forests the increment of wood stops when the share of leaves of the total mass is below 1%. At the same time, because the proportion of detritus production increases, increment of biomass becomes even slower than that of the net primary production. When the old age phase is reached, the rate of *detritus production* is higher than the rate of *net primary production* and the *biomass* decreases. Any of these changes can be modified or stopped when a stand has trees of various ages and aging and rejuvenation are simultaneous. A description of the age dependence of the available supply and turnover for a particular ecosystem can only be interpreted when the age of the stands is known.

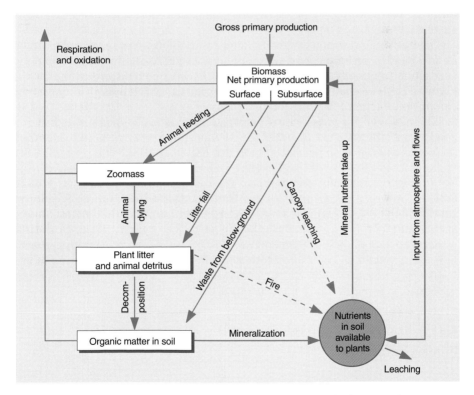

Fig. 5.4. Base model of a natural or near natural ecosystem as applied to typical ecosystems of some ecozones (e.g. Fig. 7.18, 8.12). The areas of the boxes and widths of the arrows correspond to the volume of supply and turnover. The circles represent the availability of nutrients to the plants in the soil, largely the exchangeable nutrients. The area of the circle represents the approximate relationship of nutrient availability in different ecosystems

Close to stationary states of some duration appear when the plant cover reaches a stage of late maturity. By far the largest proportion of the temporal aging cycle and also the spatial distribution within an ecosystem is accounted for by this stage. For this reason conditions during this mature stage are regarded as a steady state and more or less constant and typical for the ecozone or ecosystem. It is also during this stage that the gains and losses in the individual components of the system are in balance for a period of time.

Figure 5.4 shows a zonal ecosystem model based on such an assumption of steady state. It is applicable to several ecozones, not necessarily for the mean states but for the characteristics of the biome or ecosystem in the zone. Clearly it provides only a framework within which the supplies and turnover rates change in the long term as the components in the system develop.

5.3
Available supply of organic matter in the ecosystem

Box 3 shows the various types of organic available supply and the turnovers in an ecosystem.

The *biomass* is the total weight of living components in the ecosystem, i. e. of all producers, consumers and decomposers. In practice, however, *biomass* is used quite frequently as a synonym for *plant mass* or phytomass because in many ecosystems the *zoomass* which is the total mass of all living animals often composes less than 1% of the total biomass.

Living plants can include dead material such as bark, structural tissue or dead roots still attached to the plant. If this *standing dead* is not deducted the term *standing crop* instead of biomass has to be used. The *standing* also includes dead trees that are still upright in a woodland.

Litter and *humus* are composed of dead organic matter lying at or near the soil surface. Soil scientists define litter as part of the humus. Ecologists subdivide the dead organic matter into litter and humus, although not always at the same place. For example, in addition to the organic matter in the Ah mineral horizon, that in the Oh horizon is often considered part of the humus. Boundaries between the O horizons cannot be well defined since most are transitional areas. In this book the entire surface layer, including the Oh horizon is defined as litter. Peat is defined as humus.

5.4
Primary production

5.4.1
Photosynthesis and respiration

Each natural ecosystem begins with a fixing of solar energy in the form of latent chemical energy, the primary energy input, by green, autotrophic plants through the process of photosynthesis. *Photosynthesis* is the production of

Selected organic available supplies and turnovers in an ecosystem
3

Available supplies

Biomass	Shoot mass (above surface)	Share of photo-synthetically active plant components
– Phytomass	Root mass (below surface)	Share of plant components that only respire

– Zoomass

Litter (L and O horizons)

Humus in soil (Ah and H horizons)

Turnovers

Gross primary production (GPP)

Net primary production (NPP)

Animal consumption and secondary production

Litter production (litter fall)

Below-ground waste production

Litter decomposition

Humification

Mineralization

Duration and rate of turnovers

Duration and rate of turnover of vegetation (productivity)	$\dfrac{\text{Biomass}}{\text{NPP}}$ (years)	or	$\dfrac{\text{NPP} \times 100}{\text{Biomass}}$ (%)
Life span of leaves	$\dfrac{\text{Leaf mass}}{\text{NPP (leaves)}}$ (years)	or	$\dfrac{\text{Biomass}}{\text{Leaf drop}}$ (years)
Duration and rate of decomposition	$\dfrac{\text{Litter supply}}{\text{Litter delivery}}$ (years)	or	$\dfrac{\text{Litter delivery}}{\text{Litter supply}}$ (%)

Duration of turnover = mean period during which a biomass, zoomass, litter or humus is exchanged completely, that is, the input and output cumulate to the total current supply.

Turnover rates = proportion (%) of a supply exchanged within a time unit e.g. mean annual gain by primary or secondary production, litter fall and humification, or loss by animal consumption, animals dying and decomposition.

Turnover duration and rates are measured by flows between individual components of the available supply or total flows between primary production and mineralization.

Note: Units of measure used are tons of dry matter (dry weight of the biomass or dead organic matter, dried at 105 °C) and kilo joules (for energy content and flows; energy equivalent of plant matter estimated at 18 kJ ≈ 4.3 kcal, per 1 gram of dry matter) per hectare and per year. Carbon content of biomass assumed at 45%, of litter about 50% and humus up to 58%. Carbon content increases during decomposition.

carbohydrates from water and carbon dioxide which are used as elements for further synthesis. Photosynthesis can be expressed as follows:

$$6\,CO_2 + 6\,H_2O + 2{,}897\,kJ \rightarrow C_6H_{12}O_6\,(glucose) + 6\,O_2 \uparrow$$

A part of the carbohydrate production is lost when respiration takes place and carbon dioxide is released. Respiration is expressed as follows:

$$C_6H_{12}O_6 + 6\,O_2 \rightarrow 6\,H_2O + 6\,CO_2 \uparrow +2{,}826\,kJ$$

The gross photosynthesis less respiration equals net photosynthesis. The remaining surplus of material and energy which is produced in the plant cover of an ecosystem by means of the net photosynthesis is the *net primary production* (NPP).

The intake of CO_2 and output of O_2 that occurs in photosynthesis and the intake of O_2 and output of CO_2 that occurs in respiration take place in the *stomata (pores)* of the leaf surface. Most of the stomata are on the underside of the leaf. In the *transpiration* process the water moves from the roots upwards in the plant and is also discharged through the pores. During a water shortage pores may be closed or narrowed so that photosynthesis is reduced or even ceases.

5.4.2
Primary production from plant stands

The *production capacity* of a plant cover, the crop growth rate, is calculated from the net yield of organic matter in *dry weight* or by its *carbon content,* per time unit (mostly per year, month or growing season) in relation to the ground area (ha or m²). Included are increases in the living biomass resulting from plant growth or, usually only in older stands, decreases (ΔB) and the losses

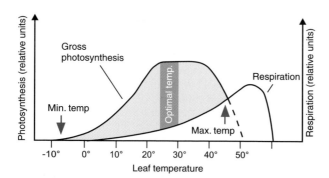

Fig. 5.5. Photosynthesis and respiration in relation to temperature. The net photosynthesis (*blue area*) is the difference between gross photosynthesis and respiration. In this example the highest value is reached between 25 °C and 30 °C. Values are higher if the light intensity increases. Source: Larcher 1994

through death (L) herbivory (C) or fire (F). The net primary production can therefore be expressed as

$$PP_N = \Delta B + L + C + F .$$

In some ecosystems, losses due to herbivory and fire are generally (or, with respect to fire, at least for longer periods) of very minor importance; the losses due to death may be negligible during "growing periods" (i.e., assuming that production and death are discrete processes). In such (simplest) cases, net primary production approximates growth increments, and the following simplified equation can be used for determining PP_N:

$$PP_N \approx \Delta B .$$

Losses due to death equal the increase in the quantity of dead plant material (ΔW), i.e., of dead vegetation (standing dead), litter and dead soil organic matter (above- and below-ground wastes) corrected for the amount decomposed (D):

$$L = \Delta W + D .$$

Substituting L in first equation, and still neglecting C and F, the following expression for determining PP_N is obtained:

$$PP_N \approx \Delta B + \Delta W + D .$$

The definition of B and W may change when live and dead vegetation are difficult to distinguish; then B takes the meaning of standing crop (i.e., phytomass and standing dead) and, correspondingly, W that of wastes only. This shift does not affect the measurements of decomposition; in any case, D has to include the decomposition processes on both the standing phytomass and the ground.

5.4.3
Production capacity of the world's plant cover

The large variations in the *primary production* per spatial unit worldwide are only to a small extent due to differences in the assimilation capacity by photosynthesis of the plant cover (Fig. 5.6). Of greater importance are size and structure of the above ground biomass, or assimilation surface if there is a great deal of photosynthetically inactive stem material present, and the extent to which edaphic and climatic conditions in a location are favorable or unfavorable for plant growth.

The *productivity* of the plant formations in the ecozones is generally higher the larger the biomass (Figs. 5.7 and 5.8). An exception are the grasslands in the steppes and grass savannas and also algae rich aquatic ecosystems which have a higher production capacity from a smaller biomass (Chap. 10).

The *assimilation surfaces* (foliage densities) and their relationship to unproductive organs is a decisive factor in the productivity of a plant cover because

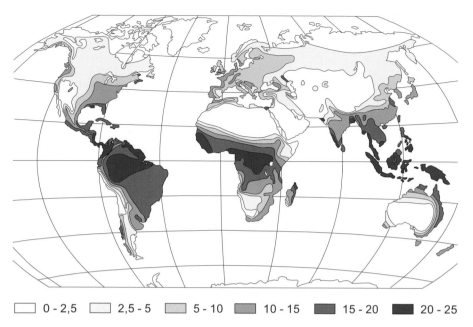

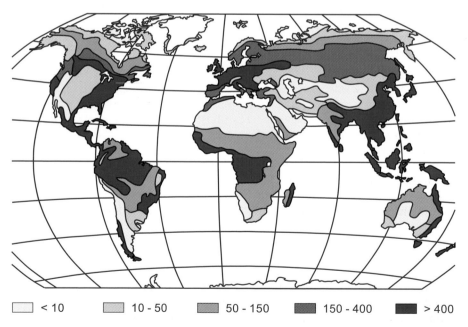

Fig. 5.6. Annual net primary production in the world in tons per hectare. Source: Lieth 1964

Fig. 5.7. Distribution of surface and subsurface biomass in the world, in tons of dry matter per hectare. Source: Bazilevich and Rodin 1971

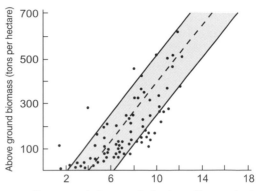

Fig. 5.8. The relationship between above ground biomass and primary production in forest formations. Source: O'Neill and De Angeles 1981

only the leaves are involved in the *assimilation of carbon*. If their total area of surfaces above ground is larger and the proportion of only respiratory organs small, other factors being equal, the productivity is also higher. This is shown by the production capacity of neighboring pastures and woodland.

The *leaf area index* (LAI) is the measure of the assimilation surface of a plant cover. This is the total of all leaf surfaces exposed to incoming light energy (projected leaf area) in relation to the ground surface areas beneath the plant. It is expressed as m^2 per $1\,m^2$ ($m^2\,m^{-2}$) and is basically a measure of light absorption and thereby the energy input of the leaves as well as the decrease of light intensity in the stand. It can also be used to estimate transpiration and interception of rainfall in a plant cover; both usually increase with increasing LAI.

In the deciduous forests of the midlatitudes the index lies between about 5 and 6, in the rainforest of the subtropics from 7 to 8 and in the rainforests of the tropics between 9 and 10. The smallest values are in the dry areas. Production increases with increasing biomass and LAI as long as the radiation interception of green plants increases.

The extent to which primary production world wide is affected by environmental factors depends on the annual length of the growing season, the solar radiation and air temperature and availability of water and mineral nutrients during the growing season.

Growing season, solar radiation, temperature and water availability are determined by climatic factors. The availability of mineral nutrients is also influenced by the climate. As a result of which productivity of the vegetation predominantly changes in relation to latitudinal changes of climate. The highest rates of productivity occur in the tropical rainforests in which the year-round rain, high precipitation totals, high temperatures and intense solar radiation combine to provide optimal conditions for growth.

Solar radiation is the most important source of energy for primary production. It is available in varying quantities world wide. Table 5.2 shows the available range for each ecozone (see also Fig. 2.1 and 2.2). The vegetation

Table 5.2. Global radiation and primary production in the ecozones

Ecozone	Global radiation during a growing season		Net primary production		
	10^8 kJ ha^{-1}	% annual total	Energy fixation 10^8 kJ ha^{-1} a^{-1} [a]	Dry weight t ha^{-1} a^{-1} [a]	Dry weight t ha^{-1} a^{-1} [b]
Polar subpolar	50–150[c]	20–50	0.25–0.75	1-4	1-4
Boreal	150–300	50–75	0.75–1.50	4–8	4–8
Temperate midlatitudes	300–400	75–80	1.50–2.00	8–11	8–13
Dry mid-latitudes	150–300[d]	25–50	0.75–1.50	4–8	4–10 (3–8)
Subtropics with winter rain	200–300	30–55	1.00–1.50	5–8	6–10
Subtropics with year-round rain	500–600	100	2.50–3.00	14–17	19–23
Dry tropics and subtropics	200–350[e] 100–200[f]	25–50 15–30	1.00–1.75 0.50–1.00	5–10 3–5	7–14 (6–11) 4–6 (3–5)
Tropics with summer rain	350–550	50–85	1.75–2.75	10–15	14–21
Tropics with year-round rain	500–650	100	2.50–3.25	14–18	21–29

[a] A mean photosynthetic efficiency of primary production is assumed to = 0.5% of the global radiation during growing season and energy equivalent of the produced plant mass to = 18 kJ g^{-1} dry matter (dry weight including mineral components).

[b] Photosynthetic efficiency increases equatorwards from 0.4 to 0.8%. Values in parenthesis are for dry areas where a reduced amount of radiation is absorbed because of the incomplete vegetation cover.

[c] Tundra   [d] Grass steppes   [e] Tropical thorn savanna   [f] Subtropical steppe

can use only that part of the solar radiation that reaches the surface during the *growing season*, the annual period or periods when thermal and moisture conditions are suitable for growth. That part of the solar radiation received at the surface at this time forms the growth potential for plant cover (Table 5.2, col. 1 and 2).

The utilization of this potential for the production of organic matter depends on the *exploitation of radiation by (apparent) photosynthesis*, i.e. on its efficiency in transforming radiation energy into chemical (biologically useful) energy. This transformation, which is variously called the energy yield or useful effect of photosynthesis, or the *(net) photosynthetic efficiency*, can be calculated in different ways. In this book it will be expressed as the *relation between the annual primary production (or, more exactly, its energy content) and the incoming radiation during the growing season.*

On the basis of the values for productivity of zonal plant formations, a global mean value for the continents of about 0.5% can be estimated, or 1% of PAR (photosynthetically active radiation). These are long term utilizaton rates of zonal plant formation, taking into account lower production related to the aging of stands. Using this value and assuming that the mean energy content of the plant matter is $18 \, kJ \, g^{-1}$, the order of magnitude of primary production for each ecozone can be estimated (Table 5.2, col. 3 and 4).

As the net photosynthesis increases with temperature equatorwards, so also does the energy yield of the NPP of the zonal plant formations from about 0.4% in the high latitudes to 0.8% at the equator, The range is probably due to the lower albedo in regions with a high sun angle. Using these zonal variations, an adjusted value for the production can be obtained (Table 5.2, col. 5).

If the water and nutrient supplies decrease, plant production is likely to be reduced. The moisture content and nutrient content at which a reduction occurs, and the extent, depends on the efficiency of the plants involved to utilize the available water and nutrients. This can be expressed as the relationship between production and water consumption and between production and nutrient uptake from the soil.

The relationship between production and water use of a single plant can be expressed by the *water use efficiency* (WUE) in grams of dry matter, or net CO_2 uptake per kilogram (liter) of transpired water. The reciprocal quotient, the *transpiration coefficient* is also sometimes used. This expresses the transpiration in liters per kilo of dry matter production.

The *water use efficiency* of the primary production from a plant cover is determined from the ratio of the production, usually above ground production, of a plant covered area in $kg \, ha^{-1}$ or $g \, m^{-2}$, to the actual evaporationtranspiration (Et) of that area, related to a time period, usually the growing season or a year.

In dry regions it can be assumed that all precipitation is evaporated where it falls and can be equated with the total mean annual precipitation and termed *rain use efficiency* (RUE) of primary production.

Water use efficiency increases with the actual evapotranspiration (Et). For example, with 100 mm Et, there is a production of dry matter of $0.05 \, g \, m^{-2}$ per mm of Et and with 1,000 mm Et, $2.0 \, g \, m^{-2}$ of dry matter per mm of Et (Fig. 5.9).

The increase in production with increasing Et relates to the supply of water to the plant cover. With a larger supply, the vegetation is taller and denser. Also, as the availability of water increases, the proportion of the evaporation that takes place at the soil surface, unused by plants, declines. The correlation between net primary production and precipitation has been measured in several plant cover types (Figs. 10.5, 13.2, 14.10).

The *nutrient use efficiency* (NUE) shows the ratio between production of a plant cover and the uptake of minerals from the soil over a time period. It is expressed as dry matter in kg of NPP of mineral uptake. The NUE is lower for herbaceous plants and leaves that have a short life span and rapid turnover and consequently short mineral cycle than woody plants which have a longer life

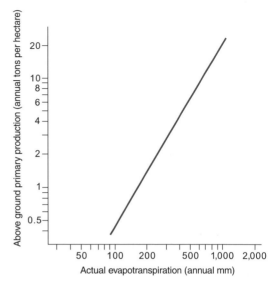

Fig. 5.9. Correlation between surface primary production and evapotranspiration, based on data from various formation types. Source: Rosenzweig 1968

span. It is also greater in soil where the mineral supply is better. The nutrient use efficiency for the grass savanna, for example is less than for the savanna woodland which is less than for tropical rainforest; similarly, in dry savanna rich in nutrients (eutropic) the NUE is less than in moist nutrient poor savanna (dystropic) and for deciduous foliage, it is less than for evergreen broadleaf trees and, in turn, less than for needles from conifers.

The nutrient use efficiency can also be used to estimate individual nutrients, nitrogen use efficiency for example. The NUE is of greatest importance in regions that are humid throughout the year. Where the water supply is at least temporarily at a minimum, the water use efficiency plays a greater role in the production capacity of the vegetation.

5.5
Consumption by animals and secondary production

Animals are heterotrophic life forms, *consumers*, that feed directly or indirectly on the organic products of the *primary producers*. They are, therefore, *secondary producers* and their products are secondary products. Depending on their feeding habits, they are classified as herbivores, carnivores, omnivores, or detrivores. The first three are biophages, the consumers that eat plants and animals. *Detrivores* consume dead matter and are often classified as *decomposers*, together with the *saprophytes* which consist mainly of bacteria and fungi, in so far as they belong to the mesofauna and microfauna of the soil up to 2 mm in size (Fig. 5.10).

Quantitatively the importance of the biophages in the ecosystem is small. Only a few percent, at most 10 to 20% of the surface biomass, is consumed by herbivores (Remmert 1992). The highest shares are in the steppes and

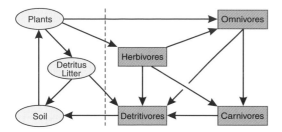

Fig. 5.10. Four main groups of heterotrophic organisms (*blue areas*). The arrows show the paths of organic material transmitted as a result of feeding by heterotrophic organisms

the savannas, where there are large animal populations, the smallest in the oligiotrophic peats of the Boreal zone (Chap. 8). Most of the primary production goes directly to the saprophages.

The material and energy flows through animal organisms is shown in Box 4. The food intake includes a smaller or larger part of the available food, the remainder is disregarded food. During digestion, part of the intake is returned unused as feces, the remainder is assimilated and becomes the gross production. This forms the total of materials and energy available for growth and reproduction by the animal (the net production) and respiration.

The turnover rate varies in speed and efficiency from one animal species to another. The efficiency of animal consumption of different species is shown in Table 14.2. Warm blooded animals assimilate 80 to 90% of the energy intake, many cold blooded animals only from 20 to 30% if they are herbivores. A large share of the energy assimilated by the warm blooded animals is used to maintain body temperature. That part of the assimilated matter which goes into net production is proportionately smaller than in the cold blooded animals. In general about 10% of the food consumed by animals is used for secondary production.

5.6
Waste and decomposition

The above ground litterfall of dead leaves, shoots, branches etc. forms a litter layer on the ground. The thickness varies around a mean value which is determined by the rate at which litter is supplied and decomposed, as long as herbivores, fire or decay of standing dead do not play an important role. Rates at which litter is added and decomposed vary greatly in different ecozones, as does the total amount of litter.

The *decomposition* of litter and below ground waste includes processes of mechanical breakdown, chemical decomposition (dissimilation) and their transition into mineral soil. The soil fauna are important in all three processes, the soil flora also in the chemical process. Many of the detrivores (detritus feeders, saprovores) and saprophytes are microscopic and include algae, fungi, actinomycetales, bacteria and numerous animal organisms.

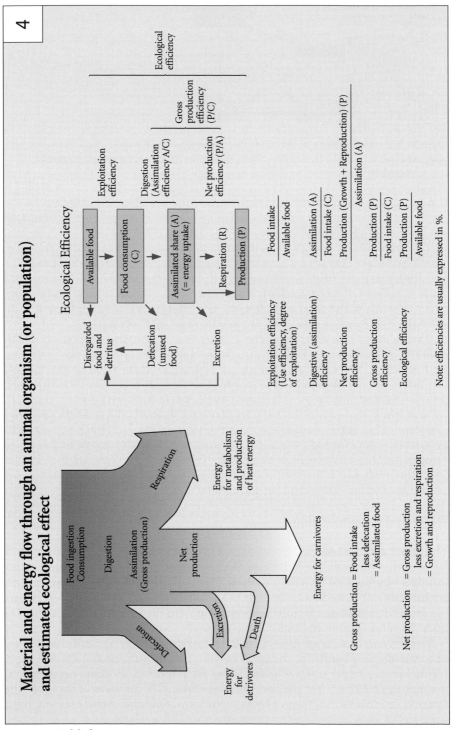

Material and energy flow through an animal organism (or population) and estimated ecological effect

Source: Ricklefs 1990

Table 5.3. Decomposition rates of broadleaf and needle litter in selected ecozones

Ecozone	Decomposition rate (k) Annual litter input / Litter accumulation	Duration of decomposition ($3/k$) Years until 95% decomposed
Polar subpolar Tundra	0.03	100
Boreal	0.21	14
Temperate midlatitudes	0.77	4
Dry midlatitudes grass steppes	1.5	2
Tropics with summer rain	3.2	1
Tropics with year-round rain	6.0	0.5

Source: Swift 1979, see also Olsen 1963

The *rate of decomposition* depends on the composition of the detritus, whether it is coarse or fine, the chemical and biological properties, the climate and edphaic factors. The rate is lower when coarse, woody components dominate in the detritus, especially if they have a high lignin content, and there are high shares of cutin, tannin and wax. Additional factors slowing the rate are low mineral content, especially if there are unfavorable C/N (carbon/nitrogen) or C/P (carbon/phosphorus) ratios, dryness, cold, stagnant water or soil acidity. Favorable C/N-ratios lie between 10:1 and 30:1; wood has a ratio of 100:1 (Larcher 1994). Limiting conditions prevail largely in cold or dry regions. The highest decompositon rates are in the tropical rainforests where humidity and temperature are high (Fig. 4.2). An annual decomposition rate can be estimated from the relationship between the annual supply of waste and the mean stock of available waste (Table 5.3) (Schultz 2000a).

5.7
Turnover of minerals

For the various biochemical synthesis following the photosynthesis, plants require several *mineral nutrients* from the soil. Nitrogen, potassium, calcium, magnesium, phosphorus and sulphur are macro elements with from 0.2–2% of dry matter. Trace elements with a content of at most 0.02% of dry matter include boron, molybdenum, chlorine, iron, manganese, zinc, and copper and, in some plants, silicon and sodium.

The amount and composition of the nutrient requirement for plants is determined by the magnitude of the PP_N and by the mineral contents of the organic substances produced (Box 5). The incorporation of mineral nutrients varies according to the species and also with the various parts of the plant.

The occurrence and flows of mineral nutrients   5

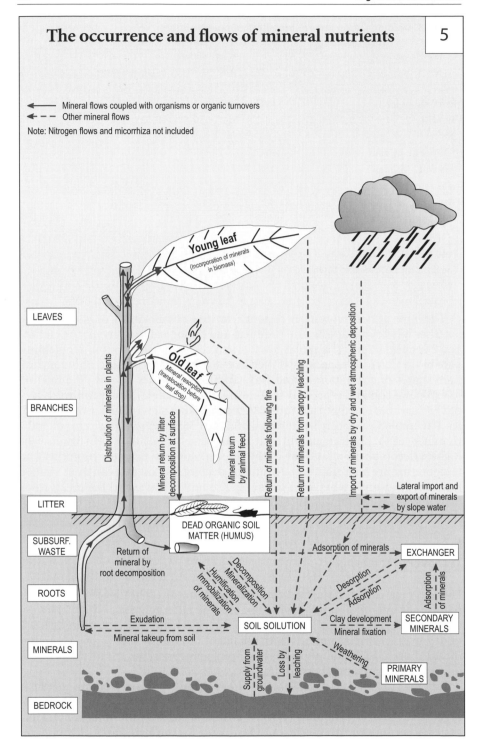

Mineral flows coupled with organisms or organic turnovers
Other mineral flows
Note: Nitrogen flows and micorrhiza not included

Young leaf
(Incorporation of minerals in biomass)

LEAVES

Old leaf
(Mineral resorption (translocation before leaf drop)

BRANCHES

Distribution of minerals in plants

Mineral return by litter decomposition at surface

Mineral return by animal feed

Return of minerals following fire

Return of minerals from canopy leaching

Import of minerals by dry and wet atmospheric deposition

LITTER

Lateral import and export of minerals by slope water

DEAD ORGANIC SOIL MATTER (HUMUS)

SUBSURF. WASTE

Return of mineral by root decomposition

Adsorption of minerals

EXCHANGER

Decomposition
Mineralization
Humification
Immobilization of minerals

Desorption
Adsorption

Adsorption of minerals

ROOTS

Exudation

Mineral takeup from soil

SOIL SOILUTION

Clay development
Mineral fixation

SECONDARY MINERALS

MINERALS

Supply from groundwater

Loss by leaching

Weathering

PRIMARY MINERALS

BEDROCK

Leaves and needles have a much higher mineral nutrient content than stems and roots. The minerals reach most higher plants with the water from the roots and are, therefore, tied to transpiration on the shoots. If drought or other stresses occur stomata may close and mineral uptake is lower. The uptake of minerals can exceed or fall short of the net primary production requirements. It can exceed them if the nutrients on the shoot or foliage surfaces are washed out in mineral form and during rainfall returned to the soil as *drops* or *trunk flow*. This is termed *canopy* or *foliage leaching*. The uptake can also exceed NPP needs if cell sap is enriched in the leaves during the growing season but not used in the growth process. The uptake can fall short of the requirements for the

---

Box 5 continued

The mineral reqirements for primary production are usually from two sources: uptake from the soil and internal recycling (the return of minerals by resorption and retranslocation from leaves before they fall). The release of minerals into and on to the soil from litter, dead animals and dying roots in an organically bound form is the result of biological and chemical decomposition processes.

The time span required for mineralization ranges from less than 1 year to several hundred years. Minerals returned by canopy leaching or precipitation are available to the forest for immediate use, as are minerals from plant internal recycling. Indirect recycling which takes place in the middle and long term and direct recycling which takes place in the short term can therefore be separately estimated. Mineral budgets in plant covers and ecosystems are usually estimated in kilograms per hectare and year. If it is assumed that the return of minerals occurs mostly by means of the surface and subsurface waste, the following relationships are valid.

Mineral uptake from soil = minerals incorporated in biomass

+ minerals returned from canopy leaching

Mineral incorporation in biomass

$$= \frac{\text{Minerals}}{\text{in biomass increment}} + \frac{\text{Mineral return by litter fall}}{\text{and below ground waste}}$$

$$\frac{\text{Turnover}}{\text{of minerals}} = \frac{\begin{array}{c}\text{Mineral return by litter fall and below ground waste}\\ + \text{Mineral return from canopy leaching}\end{array}}{\text{Mineral uptake from soil}}$$

$$\text{Mineral use efficiency} = \frac{\begin{array}{c}\text{Minerals in biomass increment}\\ + \text{Mineral return by litter fall}\\ \text{and below ground waste}\end{array}}{\text{Mineral uptake from soil}}$$

NPP if minerals incorporated in the organic matter of the leaves or needles are translocated to the shoots or roots that do not drop or die off in the autumn or at the beginning of the dry season. These minerals are, therefore, available to the net primary production during the following years. The share of translocated nutrients can be estimated by measuring the difference in minerals between mature and old leaves or recently fallen leaves, taking into account the effects, if any, of leaching. Nitrogen and potassium are often involved in the translocation process.

The return of minerals to the soil takes place in mineral form following canopy leaching or fire and in organic form in the litter, subsurface waste or animal feed. For the release of organically bound nutrients from plant waste and animal remains there are varying time spans depending on their composition and the rate of decomposition of organic matter. Potassium, for example, is released more rapidly than either nitrogen or phosphorus. Where rates of mineralization are low, the release of both nitrogen and phosphorus is at a minimum with a limiting effect for plant growth.

CHAPTER 6

Land use

The use of the land by man has lead to a widespread transformation of the original natural landscape (Fig. 6.1). The ecozonal differentiation has, however, been preserved despite the pattern of land use and the alteration of some of the original characteristics. Many forms of agricultural and forest land use reflect decisions made by man over, historically, very long periods of time. Most of these decisions harmonized closely with the possibilities offered by the natural conditions, or have, at least, adjusted to them. Much of the natural vegetation has consequently been replaced by an agricultural landscape which is adapted to the pre-existing natural conditions but which still reflects closely the original ecozonal differentiation and its subdivisions.

The natural environments of the ecozones determine both qualitatively and quantitatively the agricultural and forest land use because the boundaries of possible types of use are set in the framework of the natural production potential of an area (Table 6.1 and Fig. 6.2). The ecozone's climate and soils also determine the way in which an increase in growth in an area under cultivation can be achieved that is greater than the net primary production of

Fig. 6.1. Remaining areas of wilderness in the world (1988). Source: McCloskey and Spalding 1989

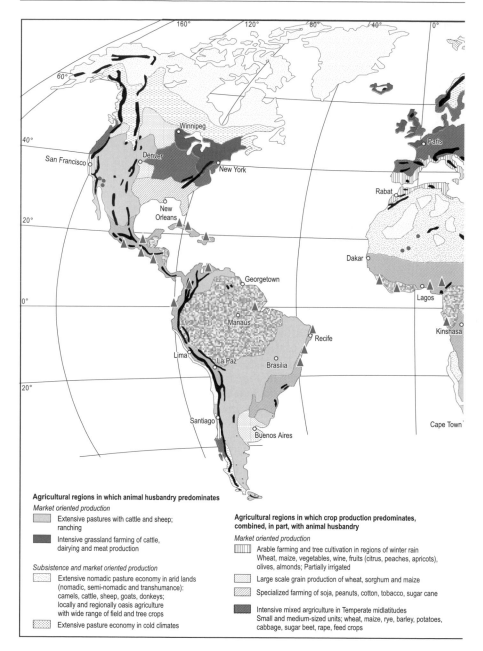

Fig. 6.2. Agricultural regions of the world. Source: World Atlas of Agriculture and other sources, including regional sources

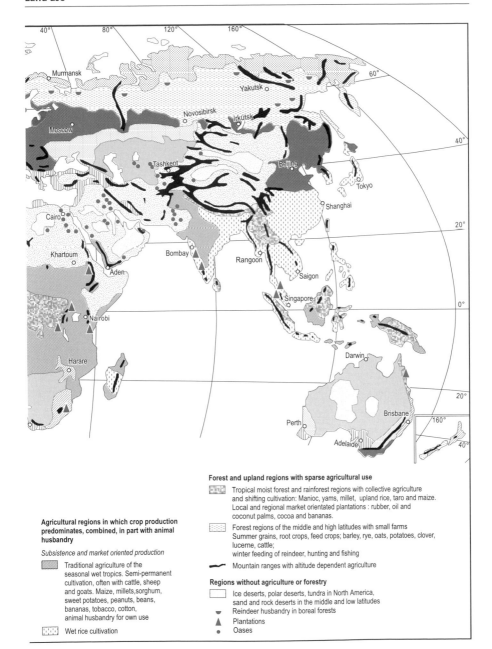

Agricultural regions in which crop production predominates, combined, in part with animal husbandry

Subsistence and market oriented production

Traditional agriculture of the seasonal wet tropics. Semi-permanent cultivation, often with cattle, sheep and goats. Maize, millets,sorghum, sweet potatoes, peanuts, beans, bananas, tobacco, cotton, animal husbandry for own use

Wet rice cultivation

Forest and upland regions with sparse agricultural use

Tropical moist forest and rainforest regions with collective agriculture and shifting cultivation: Manioc, yams, millet, upland rice, taro and maize. Local and regional market orientated plantations : rubber, oil and coconut palms, cocoa and bananas.

Forest regions of the middle and high latitudes with small farms Summer grains, root crops, feed crops; barley, rye, oats, potatoes, clover, lucerne, cattle; winter feeding of reindeer, hunting and fishing

Mountain ranges with altitude dependent agriculture

Regions without agriculture or forestry

Ice deserts, polar deserts, tundra in North America, sand and rock deserts in the middle and low latitudes

Reindeer husbandry in boreal forests

Plantations

Oases

Table 6.1. Agricultural regions and corresponding ecozone

Agricultural region	Ecozone
Nomadic and oasis agriculture in dry areas	**Dry tropics and subtropics,** Dry midlatitudes and **Subtropics** with winter rain in Eurasia
Nomadic and semi-nomadic agriculture	Deserts and semi-deserts
Transhumance	Thorn savanna, subtropical steppes, Subtropics with winter rain
Extensive grazing in cold climates (reindeer)	**Polar subpolar** and Boreal in areas of tundra and northern taiga in Eurasia
Extensive grazing – ranching	Dry midlatitudes in the Americas; Dry tropics and subtropics in Central and South America, Australia and southern Africa; Tropics with year-round rain in South America
Intensive pasture agriculture	**Temperate midlatitudes** in coastal regions of Europe, North America and Australia
Traditional agriculture of the seasonal tropics	Tropics with summer rain, extending in some areas into the thorn savanna of the Dry tropics and subtropics and forests of the Tropics with year-round rain in Africa, India (except in areas of irrigated rice cultivation), Central and South America (except in areas of ranching)
Irrigated rice cultivation	**Tropics with summer rain,** Subtropics with year-round rain and Tropics with year-round rain in Southeast Asia
Arable and plantation and orchard agriculture	**Subtropics with winter rain**
Large scale grain farming	**Dry midlatitudes** (steppes) and Dry tropics and subtropics (steppes) in South America and Australia
Specialized crop farming	**Subtropics with year-round rain** (excluding Southeast Asia)
Intensive mixed farming in temperate areas	**Temperate midlatitudes**
Tropical moist forest and rainforest regions with gathering, shifting cultivation and plantation agriculture	**Tropics with year-round rain,** Tropics with summer rain (moist savanna); gathering also in the thorn savanna of the Dry tropics and subtropics, in southern and East Africa and Australia
Forest regions in mid- and high latitudes with small farms cultivating grains, root and feed crops	**Boreal** and Temperate midlatitudes (temperate rainforests)
Largely without any agriculture	**Polar subpolar** (ice deserts, polar deserts and tundra in North America), **Dry midlatitudes** and **Dry tropics and subtropics** (sand and rock deserts; high mountain areas).

the plant cover in the ecozone. Better use of the natural production period and its artificial extension by irrigation or production under glass are examples. The use of fertilizer, pecticides and optimally productive types of seed are also related to the natural conditions.

Much more intensive agricultural land use is only possible with much higher inputs of technology and energy, often from fossil fuels, leading to a much higher energy input in relation to the energy content of the yield from agriculture.

Regional section:
The individual ecozones

Polar subpolar zone

7.1
Distribution

The Polar subpolar zone is bipolar. Towards the equator, the boundary is formed by the *polar tree line* in the Northern Hemisphere. The entire zone lies within the area of continuous *permafrost*. It has a total area of 22 million km², 15% of the global landmass, of which 14 million km² lie in the Antarctic.

About three-quarters of the total area is ice covered, forming *polar deserts* which extend over most of the Polar zone in the Southern Hemisphere. In the Northern Hemisphere Greenland and some Arctic islands have ice caps but the area is otherwise free of ice cover.

The boundary between the ice covered areas and ice free areas coincides more or less with the *snow line*. Polewards the snow line lies where more snow falls in most years in winter than is melted in summer, and towards the equator, where almost all the winter snowfall thaws in summer. The snow

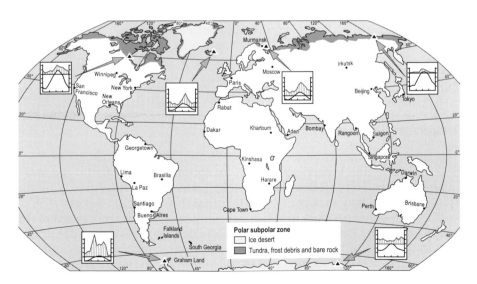

Fig. 7.1. Polar subpolar zone

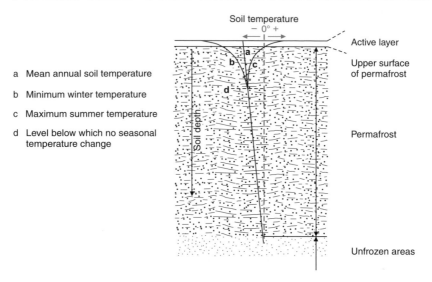

a Mean annual soil temperature

b Minimum winter temperature

c Maximum summer temperature

d Level below which no seasonal
   temperature change

Fig. 7.2. Vertical distribution of temperature in the permafrost and freeze-thaw layer. Source: Karte 1979 and Sugden 1982

and ice boundaries do not correspond because ice lobes and glacier tongues advance beyond their areas of supply into the regions that are free of snow cover.

Based on the temperature conditions and the vegetation, the ice free areas of the Polar zone can be subdivided into a zone of *frost debris and bare rock* and a *tundra* zone. In contrast to the polar ice deserts where water is present only in solid form, in the periglacial regions there is a change each year from soil ice to soil water, or from snowfall to rainfall so that each year a summer thawed layer *(active layer)* develops in the permafrost (Fig. 7.2).

7.2
Climate

7.2.1
Temperature, length of day, precipitation

The mean temperature in the warmest month in the tundra lies between 6 °C and 10 °C and for generally three, exceptionally four, months the mean temperature is > 5 °C. Polewards the high summer mean is below 6 °C in the frost debris and bare rock zone and is less than 2 °C in the ice deserts.

The drop in temperature in the winter months on the oceanic coasts of the subpolar zone is relatively small but considerable in the areas around the poles and in interiors of the continents. The difference between the highest and lowest monthly means ranges from 10 °C in subpolar maritime areas to 50 °C in the continental interiors. The diurnal temperature range is small throughout the zone.

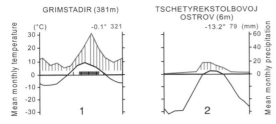

Fig. 7.3. Climates in Polar and Subpolar areas. Grimstadir (Iceland) shows data for a polar oceanic climate with small temperature amplitude and relatively high precipitation. Tschetyrekstolbovoj Ostrov (Siberia) shows data for a high polar continental station with a large temperature amplitude and low precipitation

Towards the poles, the *diurnal variations in incidence of sunlight* decrease progressively. The daily change from day to night is replaced by the semi-annual change from polar night to polar day. In line with this, the diurnal variations of the air temperature diminish so that a thermal and solar *seasonal climate* predominates. The long days and the high proportion of scattered radiation, about 50% of global radiation in the Arctic, reduces the effect of exposure more than in any other ecozone. Because of the low sun angle, the angle of slope is of much greater importance than its orientation for the maximum reception of radiation

Annual *precipitation* is low, not because of the infrequency of precipitation events, but because of their temperature-related low intensities. Apart from coastal areas, the annual totals are usually less than 200–300 mm with a large proportion falling as snow. Snow cover is rarely deeper than 20–30 cm.

7.2.2
Annual temperature changes in the soil and air layer next to the soil

In winter the snow cover protects the plants and soil beneath from the low temperatures in the atmosphere above the cover of ice and snow. In spring, however, the snow cover prevents a warming of the soil that matches the increase in air temperature. Once the snow has thawed, solar radiation falls directly on the soil surface so that the soil and the air layer directly above the soil are warmed and the soil begins to thaw. Plant life is then activated. The actual date on which the *growing season* begins varies locally by several weeks depending on the depth of the winter snow layer, which may accumulate in deep drifts, or have been blown away to leave almost snow free surfaces.

The warmed air layer immediately above the soil surface is present throughout the summer Even at the beginning of the autumn, the temperature remains above freezing, although the air only a meter above the ground level may be below the freezing point. The actual growing season generally lasts longer, therefore, than implied by climatic data from a weather station (Fig. 7.4).

The soil freezes from the surface downwards in the autumn but the subsoil above the permafrost table remains unfrozen for much longer. As the freezing

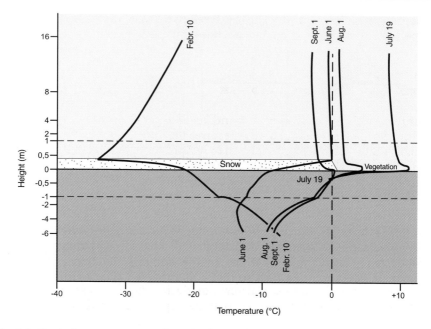

Fig. 7.4. Vertical temperature gradients in the tundra near Barrow, Alaska 71°N. The thin snow cover moderates the effect of the low winter temperature on the soil surface only slightly but delays soil warming in early summer. In the second half of the summer, the temperature in the air layer immediately above the soil is higher than in the air layers above. Scale of upper soil and air layer above the soil exaggerated. Source: Weller and Holmgren 1974

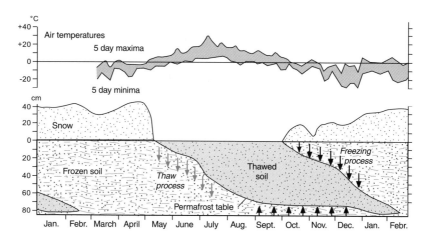

Fig. 7.5. Annual temperatures 200 cm above ground level, snow cover and freeze-thaw at Stordalen (Abisko), Sweden. From October water and air are enclosed between the permafrost and refrozen thawed area, which freezes from the surface downwards. The frost free areas are reduced in the following months but may remain until the spring in maritime areas. Source. Skartveit et al. 1975

front progresses downwards, pressure develops between the frozen surface and the permafrost which may lead to frost heaving (Fig. 7.5).

7.2.3
Summer solar radiation and heat budget

Figure 7.6 shows the *solar radiation* and *heat budget* at a station in the tundra. Worldwide, the Polar subpolar zone receives the lowest amount of solar energy annually (Table 5.2 and Fig. 2.1). From only April to September in the Northern Hemisphere and November to February in the Southern Hemisphere is the solar radiation balance positive. But, because the entire insolation is concentrated in a few months, the total daily radiation is relatively large (Fig. 2.2). In June in the Northern Hemisphere it peaks at over 60×10^8 kJ ha^{-1} mon^{-1}, similar in quantity to radiation at lower latitudes but with only half the intensity since in the polar regions it is spread over 24 hours.

The maximum warming of the land surface and atmosphere each year in summer is much less than in any other ecozones and progresses very slowly for three reasons. The radiation absorption during the first half of the summer radiation period is low because of the low angle of the sun's rays and long

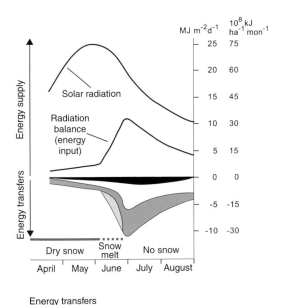

Fig. 7.6. Summer radiation and heat budget in the tundra on Axel Heiberg Island, Canada (79°N, 90°W). Source: Ohmura 1984

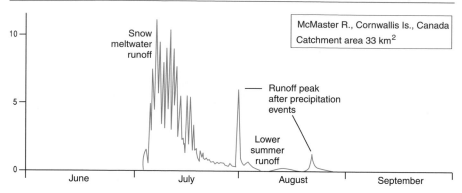

Fig. 7.7. Runoff of a stream in the Polar subpolar zone in a permafrost area. Runoff is limited to the summer months. Source: after Prowse 1994

lasting snow cover; the rate at which the soil warms up remains low because of the high heat capacity and conductivity of the soil which contains a large proportion of water and ice; and a large proportion of the energy leaves the ecosystem as latent heat in the form of sublimation and evaporation.

7.3
Relief and drainage in periglacial areas

Despite the low precipitation and consequently limited flows of water, *erosion* and *wash denudation* processes dominate in periglacial areas. From 50% to 70% of the precipitation does not infiltrate but drains off the surface, also, 80% to 90% of the total annual runoff takes place in June and July during the snow melt when most of the meltwater flows over the frozen or only partially thawed impermeable surfaces which often lie at very low slope angles. Once the surface layers are unfrozen, large quantities of fine material are removed. Stream levels rise rapidly during this period and the streams have a greatly increased erosive capacity, helped in part by ice blocks carried along in the flow. Glaciofluvial erosion can, however, only occur after the frozen valley floors, including the stream gravels, are broken through (Fig. 7.7).

Characteristic for the polar zones is the process of *freeze thaw*. The volume changes (water increases by about 10% in volume on freezing) that result from the alternating freezing and thawing of water in soil and rocks cause frost shattering, cyroturbation and gelifluction. The effectiveness of the freeze thaw process is a function of its frequency and of the water content of the regolith. The following are some of the processes and forms that result from freeze thaw.

1. *Frost shattering* can cause the break up of large blocks into rock debris of varying sizes, as well as removing blocks from the bedrock. Angular debris is characteristic of the polar deserts and northernmost areas of the Arctic tundra and forms block fields on level or gently sloping ground and debris slopes or screes at the foot of steep slopes or rockwalls.

Fig. 7.8. Types of patterned ground. Freeze-thaw processes sort material of different sizes into polygons (A and B) garlands or stone stripes (C and D). Source: Ganssen 1965

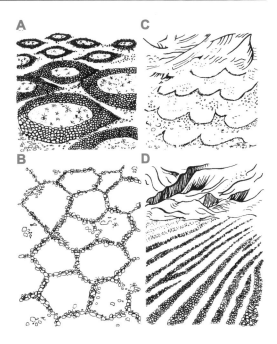

2. *Patterned ground* is the result of the vertical and horizontal sorting of originally unsorted material on level or slightly inclined surfaces that are more or less free of vegetation. At the surface this sorting process creates *stone rings*, *polygons* and *stone lines*. Stone rings and polygons can form extensive networks (Fig. 7.8)

3. *Frost wedges* are primarily caused by the contraction of ice in the permafrost at depth during the progressive cooling in winter to temperatures of at least −15 °C to −20 °C. Polygon fields then develop in which the polygons can have diameters of 10–40 m at the surface. The centers of the polygons are surrounded by heaved up ridges of surface material above the ice wedges. The lower areas of the polygons between the ridges are often water filled in summer and become polygon mires if peat forms in them (Figs. 7.9 and 7.10).

4. *Hummocks*, which are built up of fine material and have a height of at most of half a meter, are the smallest form uplifted by frost action. They occur in dense fields and have a cover of vegetation but do not have a perennial frozen core of soil or ice.

5. *Palsas* are composed of peat and are considerably larger than hummocks with heights up to 10 m. They have a core of perennially frozen substrate in which there are often ice lenses. Fields of palsas are common in the forest tundra (Chap. 8.5.3).

6. *Pingos* are the largest forms created by frost action with heights from 50 to 100 m and a large closed body of ice at the core (Fig. 7.11).

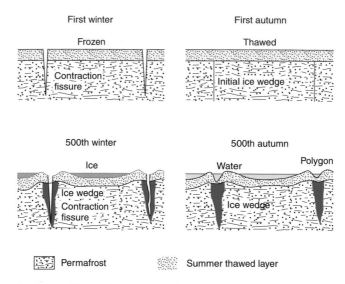

Fig. 7.9. Development of ice wedge. Fissures, developed as soil contracts in winter, fill with meltwater in summer which freezes in the following winter. Annual freeze-thaw over a long period result in the development of ice wedges which join in a polygon structure at the surface. Source: from Sugden 1982

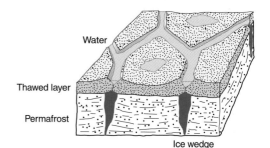

Fig. 7.10. Block diagram of ice wedge polygon in summer. Source: Butzer 1976

7. *Hollows (thermokarst)* are formed by settling following the melting of an area of permafrost that has a high ice content or of fossil ice in pingos and palsas or of an ice wedge. Such hollows are termed thermokarst because of the similarity to karst forms. Diameters of the hollows range from several hundreds to more than 1,000 m and have a depth when filled with water of 1 m, at most 4 m. When not filled with water, they can create conditions for the development of a pingo (Fig. 7.11). Hollows of this type and forms due to frost heaving can only develop if the thermal equilibrium in the permafrost is altered locally. Changes in the insulating effect of the vegetation, over grazing, stagnent conditions or burning are all possible causes. The extent of any settling depends largely on the excess ice or soil water content in the thawing permafrost. This can reach up to 40% to 50% of the total volume of the permafrost (Fig. 7.12).

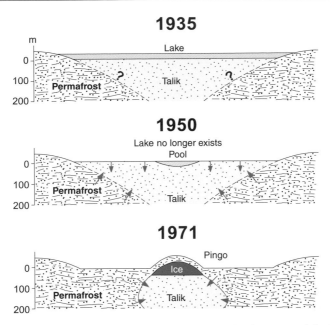

Fig. 7.11. Development of a pingo. In this example of pingo development, a lake dries up and the area of unfrozen ground (talik) begins to freeze from the sides. With the beginning of the cold season this enclosed soil water body freezes from the surface downwards, exerting pressure on the remaining water below which, however, cannot be reduced in volume and is pressed upwards. As a result the upper soil layers are pushed up and the water in them freezes as an ice lens, causing a mound, the pingo, at the surface. Source: after Mackay 1972

8. *Gelifluction* is a slow mass movement of saturated debris down slope, the result of both gravity and the expansion and contraction of the material during freeze thaw. The area affected by gelifluction in polar zones is much greater and more widespread than that covered by other periglacial forms. Gelifluction takes place when there is sufficient fine material and consequently also water storage capacity in the regolith together with a slope of at least 2° to 3°. The long term process of lowering and flattening slopes is termed cryoplanation. Periglacial surfaces can, in effect, be either denudation or accumulation forms. The effectiveness of gelifluction is less than might be suggested by the frequency of the forms created by the process. Most gelifluction surfaces are covered by loose material in which some soil development has usually taken place, indicating that the denudation rate must be slower than the low rate of weathering.

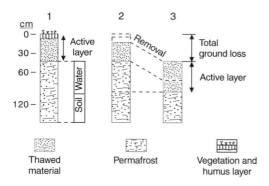

Fig. 7.12. Consequences of the destruction of a vegetated tundra soil above an area of permafrost with a high ice content. Column 1 shows the undisturbed organic layer, a 45 cm active layer (summer thawed layer) with permafrost below containing 50% excess ice (water). Column 2 shows a 15 cm reduction in the organic layer, which reduces the insulation at the surface. Column 3 shows the increase in depth of the active layer to 60 cm, of which 30 cm are in the original active layer and 30 cm developed in 60 cm of permafrost from which the 50% excess water (formerly ice) has been released, resulting in a further lowering at the surface of 30 cm. Source: French 1981

7.4
Soils

Soils in the Polar subpolar zone are termed *Cryosols*. In the tundra and zone of frost debris and bare rock the following factors are important in the formation of soils.

1. The soil formation processes are confined to a few summer months and are effective to maximally one meter in depth, unless the substrate is coarse grained, in which case thawing can occur to greater depths.

2. The permafrost layer beneath the thawed layer limits infiltration of meltwater and rainfall, leading to the formation of wetlands.

3. With an increase in soil water content, *cryoturbation* and gelifluction increase and replace soil forming processes in some areas.

4. Because of the predominance of mechanical weathering, particularly frost shattering, the soil texture is coarse grained

5. Chemical and biological turnovers are impeded. The break down of humus and release of mineral nutrients and clay mineral formation are very slow. Considerable humus enrichment can occur (Fig. 7.13).

Gelic Gleysols are the most frequent soil type in this ecozone. Below a peaty upper soil, histic (H) horizon, of 40 cm, at most 60 cm if there is a high proportion of moss, there is a relatively abrupt change to a blue-grey loamy Gley horizon, G horizon, whose upper mottled layer, the Go horizon, corresponds to

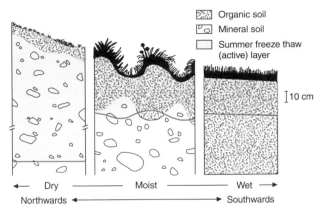

Organic soil

Mineral soil

Summer freeze thaw (active) layer

⌶10 cm

◄— Dry ————— Moist ————— Wet —►

Northwards ◄————————————► Southwards

Fig. 7.13. Characteristics of arctic tundra soil under different conditions of moisture and temperature. Organic matter (including litter) content increases with increasing soil moisture (usually in a southerly direction in the tundra) from very low in well-drained cold dry locations to very high in saturated locations in relatively warmer areas. Higher humus and litter content, denser vegetation cover, and therefore greater insulation, reduce the depth of the freeze thaw layer. Deep humus or organic layers in areas of permafrost decompose only very slowly. Source: Nadelhoffer et al. 1992

the water-logged layer which develops in summer at the bottom of the thawed layer directly above the permafrost table.

Gelic Cambisols develop in locations that remain well drained into the late summer and where soil formation is relatively advanced with several centimeters of A horizon and a dark brown Bw horizon (Arctic brown earths).

In hollows with sufficient vegetation growth *Gelic Histosols* develop. These have a thicker histic horizon than tundra gley soils. Towards the poles the formation of Histosols becomes less and less frequent and is absent altogether in the polar deserts.

Soil formation is very limited in the frost debris and bare rock areas. On level ground skeletal or coarse grained *Leptosols* or clayey or loamy *Gelic Regsols* may be present. Both have weakly developed A horizons with no raw humus or peat and gleyzation. In Canada and Greenland these soils are alkaline and in Russia mildly acid (Aleksandrova 1988), in contrast to the often high acidity of most other polar and subpolar soils.

7.5
Vegetation and fauna in the tundra and polar deserts

Very few plant species survive in the limited environment of the tundra and polar deserts. A short, cool growing season, poorly drained, nutrient poor soil, cyroturbation and gelifluction all affect the development of vegetation. In most areas fewer than 10 species make up 90% of the vascular plants (Körner 1995). In the polar deserts, there are no sedges, peat, moss or dwarf shrubs.

Most of the vascular plant species are hemicryptophytes and chamaephytes with 10% to 20% cryptophytes. Summers are too short for annual plants to complete the cycle of germination and seed formation. Chamaephytes are usually less than 30 cm high, the height of the snow cover in late winter. Because of the permafrost, root development is limited to only a few centimeters, or, in the case of crustose lichen, a few millimeters. Chamaephytes and hemicryptophytes develop in the layer of air immediately above the soil where temperatures are favorable for growth and the winter snow cover protects the shoots from extreme cold, wind shear, ice crystal formation and drought. The latter can occur during early summer heating period if the surface is bare. When growth begins the root systems are intact and, in the case of chamaephytes, there is a shoot system ready to produce leaves and blossoms for reproduction so that photosynthesis can begin. The life span of some species increases with increasing latitudes and in the high arctic many herbaceous plants hold their leaves throughout the winter.

7.5.1
Distribution of vegetation

In the southern tundra of the Northern Hemisphere the plant cover is closed. Towards the pole with increasing cold and, in summer, aridity, plant cover is generally reduced to a few favorable locations. The periglacial regions can be divided into circumpolar zones based on the the coverage of vascular plants: the *low arctic tundra*, the *high arctic tundra* and the *polar desert* (Figs. 7.14 and 8.9). The boundary between the tundra and the frost debris and bare rock zone lies in the high arctic tundra (Fig. 7.15).

Small scale variations in the pattern of vegetation cover, based on slope aspect, soil and the related local heat and moisture budgets superimpose the

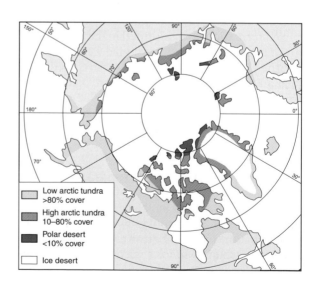

Fig. 7.14. Vegetation cover in the Polar subpolar zone. Source: Ives and Barry 1974

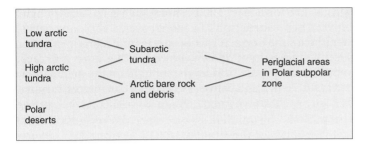

Fig. 7.15. Subdivision of the periglacial areas of the Polar subpolar zone

broad differences based on latitude. The depth and duration of snow cover on a slope affects the time at which the summer growing season begins. In addition, the degree of cold and stress caused by lack of moisture in winter and early summer plays a significant role in the initiation of plant growth in summer. Locally also, plant cover on slopes can be damaged by ice particles or the removal of fine material and, depending on aspect, solar radiation on a slope can benefit or slow down the heating up of roots and the layer of air near the soil.

Overland flow and interflow on slopes influence soil moisture, the depth of the thaw layer and the effectivenesss of frost action in the soil. The depth of the summer thaw layer is greatest on well drained mineral soils on south facing slopes (Fig. 7.13). In hollows where heat energy is added from the inflow of water, the thaw layer may be deeper.

On slopes with little snow, exposed to the wind and consequently cold and aridity, lichen and sclerophyllous evergreen dwarf shrubs dominate. Where the snow cover is of longer duration on lee slopes or in hollows, mosses are the predominant vegetation. If the snow cover is of sufficient depth but thaws early in the season, growing conditions are suitable for dwarf shrubs. In poorly drained locations, dwarf willow communities, grasses and sedges develop.

7.5.2
Biomass and primary production

Biomass increases with increasing distance from the poles and decreasing altitude, a reflection of the rise in temperature and lengthening growing season. The maximum biomass is $30\,t\,ha^{-1}$. The maximum *leaf area index* (LAI) is over 1. In the low Arctic tundra during the growing season from June to August photosynthesis continues throughout the 24 hours of daylight. Even towards the end of the growing season, the gross primary production is positive, that is, greater than the loss from respiration (Fig. 5.5). Tundra plants have low light compensation and light saturation points and behave similarly to shade plants in warmer climates. The short growing season and poor soils result in a *primary production* total of 1 to $2\,t\,ha^{-1}\,a^{-1}$, maximally 4 t, lower than any other humid region of the world. Because of the very slow growth conditions, even woody dwarf shrubs live for a least 100 perhaps 200 years. Tundra plant formations are sensitive. After destruction of a plant cover, a very long period is required for regeneration of the original mix of species and age combination (Remmert 1980). Several studies have shown, however, that after a fire, for example, a return to the former production levels, though not the original mix of species, is relatively rapid. Henry and Gunn (1991) report an interesting case. In the summer of 1987, between 500 and 1,000 caribou were isolated on a 40 km^2 island after the sea ice melted. The animals starved once the vegetation cover had been eaten. The following year the vegetation had recovered completely in terms of both the diversity of species and the production capacity, and was similar to normal tundra on the neighboring mainland.

7.5.3
Animals and animal feed

The mean ratio of the biomass converted by *herbivores*, particularly mammals and birds, is about 5% to 10% annually of the net primary production. Compared to other ecozones this is a large share, only in the steppes and savannas is the significance of herbivores greater. Herbivores are of particular importance in the tundra because the decomposition by microbes in the soil does not keep pace with the delivery waste from the vegetation. Herbivores contribute, therefore to the maintainance of the mineral cycle. The plants are dependent on the consumers and the consumers on the plants.

The most important herbivores are ungulates such as reindeer, caribou and musk ox, rodents (lemmings) and hares and rabbits. Ptarmigan and wild geese are also important.

Small species of animals are subject to extreme cyclical deviations in population in a period of a few years. They are able to react to the sudden occurrence of favorable conditions that encourage a higher rate of reproduction, in contrast to large animals which adjust more slowly and are able to survive unfavorable

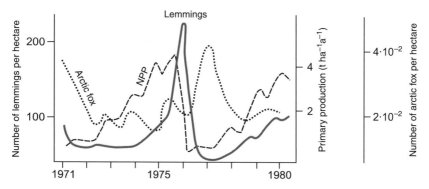

Fig. 7.16. Changes in the lemming population (Dicrostonyx torquatus) in relation to the productivity of the tundra vegetation and densities of arctic fox (Alopex lagopus) on Wrangel Island, Canada. Source: Chernov and Matveyeva 1997

conditions (Remmert 1992). *Lemmings*, for example, with a gestation period of 21 days produce up to 9 broods annually with an average of 7 young so that in favorable years there can be a rapid increase of population (Batzli 1981). During a 20 year period near Barrow, Alaska the density of lemmings rose to 150 to 225 per hectare annually every six years and fell in the intervening years to 1 to 5 animals (Bliss 1997). Variations in food supply are the usual cause of variations in density, whereby an increase or decrease in animals such as lemmings, voles and ptarmigan is often, in part, a cause of the fluctuation in food supply, The growth in population during years of plentiful food eventually leads to shortages and a reduction in population When the vegetation and, therefore, food supply recover the cycle begins again.

Vegetation recovers rapidly during a phase in which the soil has a higher available supply of minerals, despite earlier over grazing, resulting in a biomass that is relatively rich in minerals. The increase in available mineral nutrients is due to the feeding of the animals which speeds up the decomposition of organic substances and therefore has a fertilizing effect.

7.5.4
Decomposition and turnover of minerals

Even worse than the ecological disadvantages of the tundra in terms of primary production are the disadvantages associated with the decomposition of dead soil organic matter. Compared with tropical rain forests, the NPP of the tundra amounts to only one tenth and decomposition takes from 100 to 1,000 years longer (Table 7.1). Because of the low decomposition rate the *litter* layer is deep and the *humus* content of the A horizon high, especially in the low Arctic. The dead soil organic matter frequently composes over 90% of the total organic matter (biomass + litter + humus), a higher constituent amount in the, albeit relatively small, biomass, than in any other ecozone. In addition, the absolute amount of dead organic matter is 300 t to 600 t ha^{-1}, which is higher than in all other ecozones except the Boreal zone. The high means in the Boreal zone result largely from the inclusion of the frequently occurring peat bogs (Chap. 8).

Decomposition is retarded because of the lack of heat, the unfavorable carbon/nitrogen (C/N) ratio of most components in the litter and the acid and often oxygen poor environment in stagnant areas. Lack of oxygen is of greater significance than temperature.

Warming of the earth's atmosphere would have little immediate influence on the storage of carbon in the tundra soils. Indirectly there would be an effect

Table 7.1. Primary production and decomposition in the tundra and in tropical rainforests

Plant formation	Estimated	
	Primary production (t ha^{-1} a^{-1})	Duration of decomposition (years)
Tundra	2	100–1000
Tropical rainforest	20	1

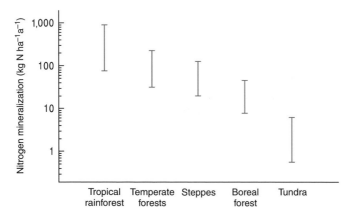

Fig. 7.17. Net mineralization rates of nitrogen within selected zonal ecosystems. Net mineralization rates are the difference between the release of nitrogen from organic compounds and the renewed immobilization by microbes (dentrification, for example). Source Nadelhoffer et al. 1992

if the summer thaw layer deepened and evaporation increased since this would be followed by a reduction of soil moisture and an increase in the volume of soil subject to decomposition.

As a result of the litter and humus enrichment, a large proportion of the nutrients cannot be taken up by plants because the accumulated organic matter forms both a *carbon sink* and a *nutrient sink*. The immobilization of nutrients particularly limits the nitrogen supply. An annual mean of only 1 to 6 kg N per hectare is mineralized, compared to 15 to 200 kg in boreal and temperate forests and 900 kg in tropical rainforests (Fig. 7.17).

Plants in the tundra adapt to the limited supply of minerals. Many dwarf shrubs are evergreen, keeping their leaves for several years and in this way reducing their time-related nutrient requirements. Their use of minerals is highly efficient and small quantities produce relatively large amounts of vegetable matter. As a consequence though, concentrations of minerals in the plant tissue are low and decomposition limited.

The deciduous chamaephytes and hemicryophytes are for the most part supplied with minerals, particularly N, K and P and carbohydrates, from reserves in the perennial shoots or roots that developed in the preceding year by translocation from dying parts. Such reserves are particularly important at the beginning of the growing season when the temperature increases but the ground is still largely frozen.

7.5.5
Model of a tundra ecosystem

The model begins with an assumed steady state, although in the case of the tundra, this is more questionable than in other ecosystems since it is possible

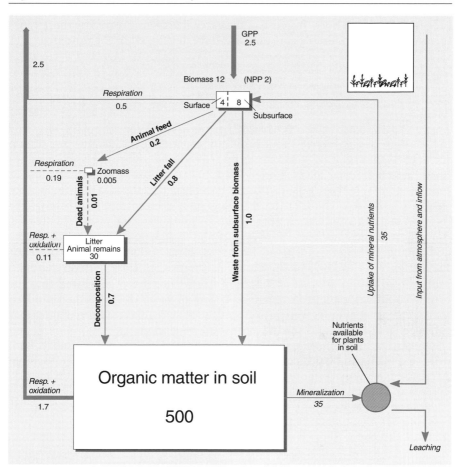

Fig. 7.18. Model of an ecosystem in a dwarf shrub tundra on a Gelic Gleysol with a Dystric H horizon. Note the small biomass and very large supply of organic matter in the soil, mostly in the form of raw humus and peat. Much of the tundra is a below ground ecosystem. Width of arrows, areas of circle and boxes are approximately in proportion to the volumes involved. Organic substances in t ha^{-1} or t ha^{-1} a^{-1}, Minerals in kg ha^{-1} a^{-1}. Source: Bliss et al. 1981; Tieszen 1978, Wielgolaski 1975, 1997, Oechel et al. 1997

that an equilibrium is never reached. At least in some places the NPP is invariably greater than the rate of decomposition so that there is a constant increase in the supply of organic matter. Studies have shown that in *peat bogs* carbon in the soils grows from 0.3 to 1.2 t ha^{-1} annually and in dwarf shrub tundra by 0.23 t ha^{-1} which would result in an increase of dry matter from 0.7 to 2.7 t and 0.5 t per hectare respectively (Chap. 8).

The worldwide total of organic carbon stored in the dead organic matter of the tundra soils is estimated at more than 50 Gt, maximally 100 Gt. The largest volumes are present in the low Arctic tussock grass-dwarf shrub tundra and sedge-lowland peat bogs. Because of the extensive areas covered by these types of vegetation, considerable carbon sinks have developed in the tundra.

7.6
Land use

There is an almost total absence of settlements in the Polar subpolar zone. Only in the subarctic tundra live about 2 million people. The indigenous population includes 90,000 Inuit in North America, Greenland and to a lesser extent in northeastern Siberia, 25,000 Samen (Lapps) in northern Europe and a larger number of people from various groups in Siberia (Samoyeds, Yakuts, Eastern Yakuts, Chukchee). The Inuit have traditionally concentrated on fishing and hunting, including the catching of seals and whales in coastal waters. In Eurasia, nomadic and semi-nomadic populations herd reindeer, grazing in the tundra and tundra-like areas of upland in the summer and moving to woodands further south or to lowlands in winter. At the present time there are about 3 million reindeer in an area of 3 million km². It has been suggested that the tundra could be developed into an important meat producing area if modern management techniques were introduced, such as regular rotation of pastures, nitrogen fertilization and sowing of productive grasses.

Domestication of musk ox might be another development possibility. Both meat and wool can be utilized and since they feed on willows and, in the high Arctic, sedges, they do not compete with reindeer or caribou which feed on grasses and lichen.

Modern settlements in the tundra have to overcome problems created by the summer thaw. Foundations of buildings must be anchored in the permafrost and at the same time not transfer heat downwards. In intense cold, the foundations of roads can be so insulated that the permafrost table rises upwards.

The supply of food and water to the population is particularly difficult. The polar limit of agriculture lies further south in the area of coniferous forests. Groundwater is frozen.

Major problems can arise from the destruction of tundra vegetation, as is common practice in the areas surrounding human settlements. Such destruction eliminates an important source of insulation for the permafrost in the soil, and as a result the thaw layer in summer penetrates increasingly deeper. If the newly thawed layer has a particularly high ice content, the soil subsides by an amount equal to the volume of excess ice. Such settling of the ground can become exacerbated when "warm" meltwater flows into the depressions formed, i.e. the process may be self-reinforcing.

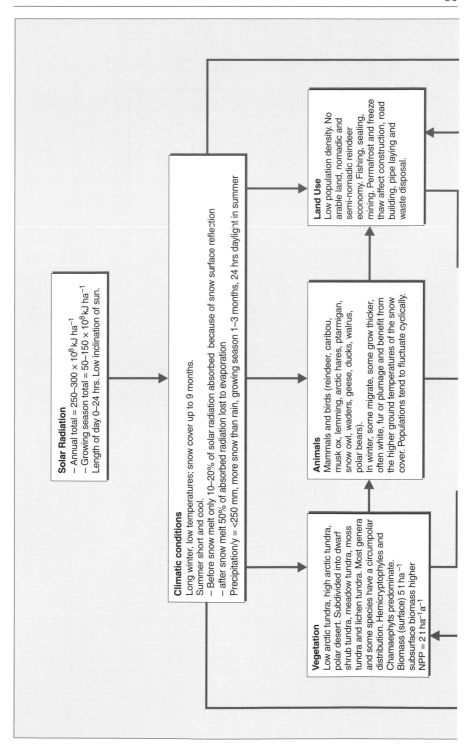

Solar Radiation

– Annual total = $250–300 \times 10^8$ kJ ha^{-1}
– Growing season total = $50–150 \times 10^8$ kJ ha^{-1}
Length of day 0–24 hrs. Low inclination of sun.

Climatic conditions

Long winter, low temperatures; snow cover up to 9 months.
Summer short and cool.
– Before snow melt only 10–20% of solar radiation absorbed because of snow surface reflection
– after snow melt 50% of absorbed radiation lost to evaporation
Precipitation/y = <250 mm, more snow than rain, growing season 1–3 months, 24 hrs daylight in summer

Vegetation

Low arctic tundra, high arctic tundra, polar desert. Subdivided into dwarf shrub tundra, meadow tundra, moss tundra and lichen tundra. Most genera and some species have a circumpolar distribution. Hemicryptophyles and Chamaephyts predominate.
Biomass (surface) 5 t ha^{-1}
subsurface biomass higher
NPP = 2 t ha^{-1}a^{-1}

Animals

Mammals and birds (reindeer, caribou, musk ox, lemming, arctic hares, ptarmigan, snow owl, waders, geese, ducks, walrus, polar bears).
In winter, some migrate, some grow thicker, often white, fur or plumage and benefit from the higher ground temperatures of the snow cover. Populations tend to fluctuate cyclically.

Land Use

Low population density. No arable land, nomadic and semi-nomadic reindeer economy. Fishing, sealing, mining. Permafrost and freeze thaw affect construction, road building, pipe laying and waste disposal.

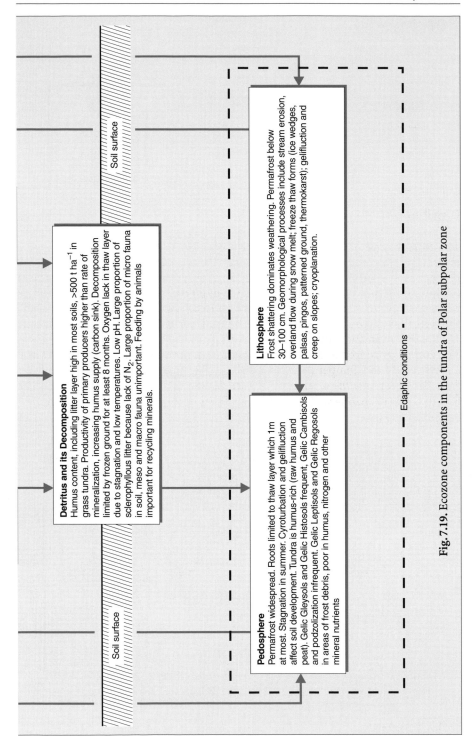

Detritus and its Decomposition
Humus content, including litter layer high in most soils, >500 t ha⁻¹ in grass tundra. Productivity of primary producers higher than rate of mineralization, increasing humus supply (carbon sink). Decomposition limited by frozen ground for at least 8 months. Oxygen lack in thaw layer due to stagnation and low temperatures. Low pH. Large proportion of sclerophyllous litter because lack of N₂. Large proportion of micro fauna in soil, meso and macro fauna unimportant. Feeding by animals important for recycling minerals.

Soil surface

Soil surface

Lithosphere
Frost shattering dominates weathering. Permafrost below 30–100 cm. Geomorphological processes include stream erosion, overland flow during snow melt; freeze thaw forms (ice wedges, palsas, pingos, patterned ground, thermokarst); gelifluction and creep on slopes; cryoplanation.

Pedosphere
Permafrost widespread. Roots limited to thaw layer which 1m at most. Stagnation in summer. Cyroturbation and gelifluction affect soil development. Tundra is humus-rich (raw humus and peat). Gelic Gleysols and Gelic Histosols frequent, Gelic Cambisols and podzolization infrequent. Gelic Leptisols and Gelic Regosols in areas of frost debris, poor in humus, nitrogen and other mineral nutrients

Edaphic conditions

Fig. 7.19. Ecozone components in the tundra of Polar subpolar zone

CHAPTER 8

Boreal zone

8.1
Distribution

The Boreal zone is the only ecozone limited to the Northern Hemisphere. It is circumpolar with a north south extent of at least 700 km; in North America it reaches a maximum of 1,500 km and in Asia 2,000 km. On the east coast of the continents its southern boundary extends to 50°N but only 60°N on western coasts because of warm ocean currents, such as the Gulf Stream and the Kiroshio. Apart from some areas of tundra, Canada, Alaska, Scandinavia, northern Russia and Siberia belong entirely or in large part to the Boreal zone. The zone is bordered to the south by steppes in the center of the continent and elsewhere by the Temperate midlatitudes. The northern boundary coincides with the *polar treeline* at about 72°N in Russia and 69°N in Canada. It is the fourth largest ecozone with a total area of almost 20 million km^2 or about 13% of the land mass.

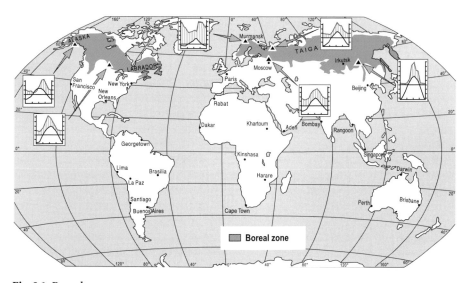

Fig. 8.1. Boreal zone

8.2
Climate

The Boreal zone has four to five, maximally six months with a mean monthly temperature of $\geq 5\,°C$. This is generally also the length of the *growing season*. Mean monthly temperatures of $\geq 10\,°C$ occur in one to three, occasionally four months. In the interiors of continents the growing season can be as short as 2 to 3 months but in all of them the mean monthly temperature is $> 10\,°C$.

The *boreal coniferous forest* is the characteristic plant formation for the zone. It is still relatively unexploited. To the south, where there is sufficient moisture and the growing season is at least 6 months and there are about 4 months in which the mean monthly temperature is $\geq 10\,°C$, the Boreal zone merges with the Temperate midlatitude zone. Elsewhere the steppes and semi-deserts of the Dry midlatitude zone form the boundary.

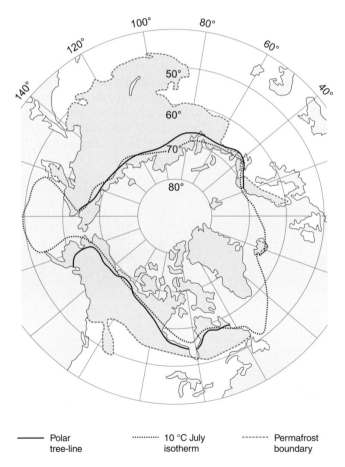

—— Polar tree-line	·········· 10 °C July isotherm	-------- Permafrost boundary

Fig. 8.2. 10 °C isotherm for July and the polar tree line in the Northern Hemisphere. Source: Stäblein 1987

To the north beyond the polar tree line, the tundra begins. The tree line lies very close to the 10 °C July isotherm (Fig. 8.2). The distribution of the permafrost and the tree line so not coincide.

The annual precipitation in the zone ranges from 250 to 500 mm, maximally up to 800 mm. Exceptions are the western and eastern coastal areas and islands of North America and Eurasia where precipitation totals can be considerably higher. Although higher than in the tundra, annual precipitation in the Boreal zone is relatively low compared to other humid zones. A large proportion of the precipitation falls as snow but the total is generally less than the total rainfall in the area. Winter snow cover ranges from 30 to 100 cm in depth, more than in the tundra, but of shorter duration, generally staying on the ground for only 6 to 7 months.

At the summer solstice there are at least 16 hours of daylight on the southern boundary of the zone and 24 hours in the most northerly areas. Similar to the tundra, the low intensity of radiation, is partially compensated for by longer hours of sunshine. The global radiation from May to July has a peak value of about 60×10^8 kJ ha^{-1} mon^{-1} or more, similar to climatic zones to the south during this period.

Air temperatures remain, however, low because a large proportion of the radiation is reflected by the extensive snow surface or transferred as latent heat by evaporating meltwater later in the season The period of higher radiation values and positive radiation balance is relatively short and the initially frozen and later, after thawing, saturated soil has a high heat capacity and conductivity and warms up only very slowly.

West east change in the degree of *continentality*, or distance from the oceans, is the reason for a regional climatic differentiation in Eurasia and, to a lesser extent, in North America. Temperature differences in both summer and winter are largely determined by the extent to which the equalizing influence of the oceans reaches into the continents. Extremes are shown in the climatic diagrams 1 and 2 in Fig. 8.3. The first showing data from a station in central

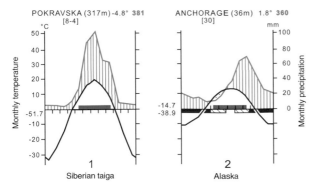

Fig. 8.3. Climates in the Boreal zone. 1. cold maritime climate 2. cold continental climate. Temperature amplitudes and the proportions of the total precipitation falling in summer increase with increasing continentality

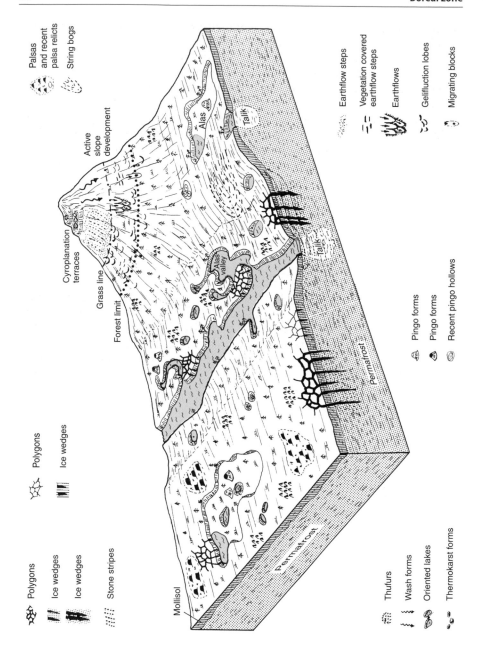

Fig. 8.4. Block diagram of landforms in a periglacial area of the Boreal zone. Source: Karte 1979

Siberia is an example of cold continental climate. Winters are very cold with absolute minima down to an extreme of −70 °C and summers warm but short with a maxima over 30 °C. The mean annual temperature is usually below −5 °C but the range of the temperatures is greater than in any other ecozone. Snowfall is low and below a certain depth the subsoil is permanently frozen.

Diagram 2 in Fig. 8.3, shows data from southern Alaska for a cold oceanic climate. Summers are cooler than further inland and winters considerably milder. The annual temperature range is consequently much smaller and the mean annual temperature higher at around 0 °C. Snow cover is deeper and permafrost is discontinuous or sporadic and in a few areas absent altogether.

8.3
Relief and drainage

Similarly to Polar subpolar ecozone, the Boreal zone was largely covered by the inland ice during the Pleistocene Ice Age. An exception was Siberia where only the central uplands and eastern mountain ranges were ice covered. The present landscape has developed, therefore, relatively recently and soils are maximally 12,000 years old. In contrast to the Temperate midlatitudes most of the Boreal zone lies in the center of the area covered by the Pleistocene ice mass. Glacial erosion processes have dominated leaving rock surfaces, roche moutonnees and rock basins, which now often form lakes. Glacial and fluvioglacial deposits are largely absent. Where they are present, they are related to material deposited as the ice retreated and melted.

As in the Polar subpolar zone, processes and forms resulting from *frost action* predominate. Large parts of the Boreal zone in Eurasia are in the area of *continuous permafrost*. In nearly all the remainder of the zone, particularly in North America, *sporadic permafrost* is widespread. String bogs, palsas, frost wedges and thermokarst are typical forms. *String bogs* develop on slopes as long narrow ridges of *peat bog* on which dwarf shrubs grow. They form a pattern of stripes or less often a network. The lower areas between the peat ridges

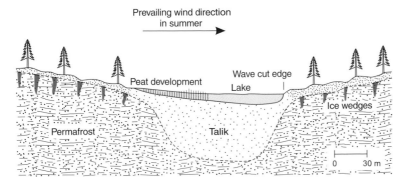

Fig. 8.5. Thermokarst lake. Subsidence at the surface is due to local more rapid thawing. Source: Butzer 1976

are usually water filled. String bogs run at right angles to the slope angle and are probably formed when the vegetation cover is torn over short stretches as a result of soil movement. Thermokarst forms extend as shallow hollows sometimes over many kilometers the result of local more intensive thawing of the permafrost, particularly if it contains a high proportion of ice, which causes loose material to collapse when water in the permafrost flows away after the ice thaws.

In sporadic or *discontinuous permafrost*, the thawing processes are frequently related to small areas of fossil permafrost. They can also occur in continuous permafrost after, for example, forest fires or clearing of the forest. Once the forest is removed the solar radiation is no longer absorbed on the crown but on the forest floor where warmer and dryer conditions are created. The permafrost thaws to a greater depth at these locations and, depending on the volume of ice in the soil, hollows form at the surface. The hollows fill with water which increases the absorption of radiation and heat flow in the soil. The thawing process is thereby accelerated and the hollows deepened. The process can be further intensified by the increased rate of decomposition of dead organic matter in the soil, formerly preserved in the permafrost. Several positive feedbacks take place in the interaction of these processes (Fig. 8.6).

If conditions are reversed lakes or ponds of this type are filled in and the newly developed vegetation cover has an insulating effect. The summer heat

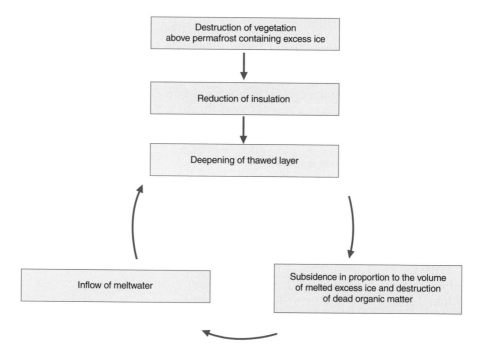

Fig. 8.6. Development of hollows following forest destruction.

flows in the soil are then reduced and large soil ice masses may develop which arch up the surface.

Stream flow in the Boreal zone is highly variable. In April or May the snow melts within a few weeks in entire stream basins, particularly if they are small, leading to extreme peaks as the streams flow simultaneously over the frozen surfaces. In the valleys, the spring meltwater flows over still frozen rivers, sometimes forming new channels. Braiding frequently follows.

After the snow melt, stream flow decreases rapidly. Summer rain, which is not large, does not contribute to groundwater storage because evaporation during the warmer months is more or less equal to precipitation. With the decrease in the air temperature and therefore also evaporation in the autumn, the surplus precipitation results in a small increase in the stream flow. As soon as the winter snow falls, this trend is reversed From now on, there is only groundwater storage flow which reaches its minimum before the snow melt in the following spring. In stream basins that lie entirely in areas of continuous permafrost, all flow ceases completely as soon as the summer thawed layer is refrozen.

8.4
Soils

Deep litter layers that are thicker than the underlying Ah horizon are characteristic of the soils in the Boreal zone. They are largely the consequence of the slow decomposition of the resinous conifer needles and the small hard leaves of many dwarf shrubs (Calluna, Vaccicium, Erica, Andromeda) in the cold and wet conditions that dominate for much of the year and which lead to the development of a highly acid litter and soil. *Peat* is present where there are stagnant conditions or the groundwater table is near the surface. Otherwise *raw humus* has developed. Both types of humus formation contain very low levels of nutrients because of the slow mineralization rates, and tend to lie at the surface, unmixed with the horizons beneath.

Podzols are widespread in the Boreal zone. They are the product of the podzolization process in which humins produced by decomposition in an acid environment are soluble and are carried down through the soil profile in seepage water to a lower level where they are precipitated out. In addition to the organic compounds, Al and Fe oxides (sesquioxides) produced by silicate weathering are also leached and deposited in the lower horizons.

With the removal of the humins and sesqioxides, a 20 to 60 cm ashy grey colored eluvial Ae or E horizon develops beneath the dark grey Ah horizon. Below the eluvial horizon is an illuvial horizon (Spodic B horizon) in which the dissolved material has been precipitated out.

The upper part of this horizon (Bh horizon) is dark brown in color because of the accumulation of humus colloids. Below, the sesquioxides give a more rusty brown color to the Bs horizon (Fig. 8.7).

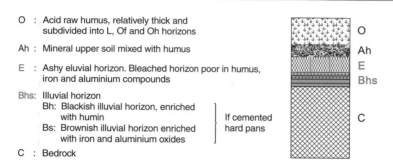

O : Acid raw humus, relatively thick and
       subdivided into L, Of and Oh horizons

Ah : Mineral upper soil mixed with humus

E : Ashy eluvial horizon. Bleached horizon poor in humus,
       iron and aluminium compounds

Bhs: Illuvial horizon
       Bh: Blackish illuvial horizon, enriched
             with humin   } If cemented
       Bs: Brownish illuvial horizon enriched   hard pans
             with iron and aluminium oxides   }
C : Bedrock

Fig. 8.7. Podzol profile

High concentrations of iron (Fe) can lead to the development of a hardened layer which usually contains a high proportion of large pores. Water can percolate in these conditions but root development may be limited. Acidity, low CEC and base saturation, coupled with a sandy A horizon in which there is little clay to prevent water percolating through and cementation in the B horizon, together underlie the low fertility of the podzols.

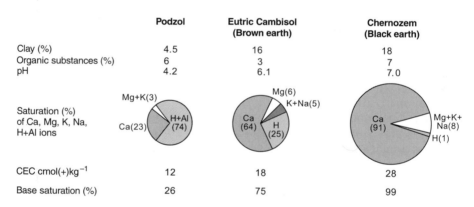

	Podzol	Eutric Cambisol (Brown earth)	Chernozem (Black earth)
Clay (%)	4.5	16	18
Organic substances (%)	6	3	7
pH	4.2	6.1	7.0
CEC cmol(+)kg⁻¹	12	18	28
Base saturation (%)	26	75	99

Fig. 8.8. Cation exchange capacity (CEC), cation composition and base saturation in a Podzol, a Eutric Cambisol (Brown earth) and a Chernozem (Black earth). The Podzol is the least favourable of the soils in all respects: its CEC is one-third smaller than that of the Cambisol and less than half that of the Chernozem; and only a quarter (26%) of the CEC is saturated with basic ions; i.e. in the Podzol only 3 cmol(+)kg⁻¹ are attributed to nutrient ions. In the case of the Cambisol, this value amounts to 13.5 cmol(+)kg⁻¹ (three-quarters of 18) and in the case of the Chernozem to 28 cmol(+)kg⁻¹ (almost the entire CEC). Cambisols possess therefore four times as many and Chernozems nine times as many nutrient cations as the Podzols. If the soil depth and the root depth are taken into account, there is an even greater difference in the availability of nutrients. Despite the relatively high share of organic matter in Podzols, 6%, compared to 7% in Chernozems and 3% in Cambisols, it is composed largely of biologically inactive raw humus in which plant remains are hardly decomposed. Source: Schroeder and Blum 1992

Histosols have developed in the Boreal zone in areas of poor drainage in the extensive lowlands of western Siberia and parts of central Canada where deep horizons of peat have been formed. They have a depth of at least 40 cm, increasing to 60 cm if the proportion of peat mosses is high. Hydromorphic soils of this type belong to the Gelic Histosols or, if they lie outside the areas of permafrost, to the Fibric Histosols.

In the more pronounced continental areas of central and eastern Siberia and the Canadian Rockies, *Cambisols* dominate, either Gelic or Dystric or more seldom, Eutric Cambisols. Gelic Leptosols occur on steep slopes. Both Leptosols and Cambisols are relatively underdeveloped soils with only a Bw horizon or AC Profile. Similar to the Podzols, fertility is low because of high acidity, lack of plant nutrients and a disadvantageous environment.

8.5
Vegetation and animals

Despite the large extent of the Boreal zone, there are few regional variations in the vegetation. Over the entire area *coniferous forests* and occasional areas of mixed forests are separated by numerous, often partially filled in, lakes and oligotrophic peat bogs. In the north of the zone, the *forest tundra* forms a transition zone with the tundra. The total number of species in the boreal regions is small although larger than in the tundra.

8.5.1
Boreal coniferous forest

Spruce (Picea), pine (Pinus), larches (Larix) and firs (Abies) cover thousands of square kilometers in single species stands. Deciduous larches extend over large areas of the interior of Siberia and form the polar tree line for all of Siberia.

Conifers replace their needles over a period of several years so that their requirements for minerals is much lower than deciduous trees which renew their foliage each spring (Chap. 9.5.4). The continuous shedding of needles over a long period is, in effect, a reflection of the poverty of minerals in the soil, the limitations on root growth because of the permafrost, the winter cold and the absence of moisture due to frost. Lack of nutrients is also expressed in the sparse stands and, compared to the spruce and fir stands of the midlatitudes, slender growth.

Unlike coniferous forests in temperate zones, a cover of shrubs and herbs is well developed. Deciduous species such as birch (Betula), poplar (Populus), willow (Salix), alder (Alnus) and ash (Fraxinus) are frequent. The herb layer includes lichens and, where stands are denser, mosses. There is also a relatively high proportion of dwarf shrubs.

The large scale differentiation of forest cover reflects the north-south differences in climate. In Russia there are three zones, the northern, central and

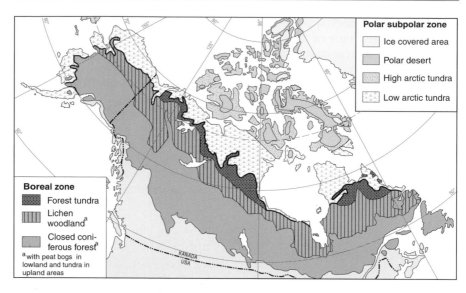

Fig. 8.9. Distribution and subdivisions of the Polar subpolar zone and Boreal zone in North America. Source: Elliot-Fisk 1989 and Bliss et al. 1981

southern *taiga*, and in North America, two: a northern area of open lichen woodland and a southern zone of closed stands of woodland (Fig. 8.9).

Small scale differentiation in the boreal forests can relate to variations in soil moisture or to exposure to solar radiation but more commonly it represents different stages of regeneration. The renewed growth in woodlands that follows events such as local forest fires, damage by insects, wind destruction or flooding results in a physiognomic and to some extent also floral mosaic composed of rejuvenation, mature and aging phases. A steady state in the sense of a forest preserved over a long period does not occur. While the stands of different age constantly change their position within the vegetation mosaic, the overall ecosystem remains stable for thousands of years. Age related changes in plant cover take place rapidly at the beginning of a regeneration period but then slow down so that the later stages of plant cover tend to be dominant in a forest mosaic.

8.5.2
Peat bogs

Peat bog covers at least 10% of the surface in most parts of the Boreal zone and in a few areas a considerably higher proportion. Trees are absent and plant life is confined to a very few species of moss, sedges (Carex, Eriophorum) herbs and dwarf shrubs. The ability of sphagnum mosses, the main component of peat, to store water allows the bog to develop above the original stagnant water level. Deposition from the atmosphere is the only source of minerals in a peat bog. Bog convexity increases with the favorability of conditions for peat bog development. Maritime climatic conditions are more suitable than the interiors of continents where in general summers are too warm and dry.

The accumulation of *carbon* stored in the post Ice Age boreal and subarctic bogs has been estimated at 455 Gt which is approximately one quarter of world wide organic carbon (Gorham 1991). This is based on a total peat area of 3.42 million km^2, a mean peat thickness of 2.3 m, a density of 0.122 $g cm^{-1}$ and a carbon content of 51.7%. The mean growth rate in the post Ice Age period, the *net sink*, has been estimated at 0.096 Gt^{-1}. At the present time the growth rate is estimated to be 0.076 $Gt a^{-1}$, or 23 $g a^{-1}$ per square meter of peat bog, an annual growth of peat (dry weight) of approximately 0.5 t per hectare (Fig. 8.13).

8.5.3
Forest tundra, polar and forest tree lines

Forest tundra is a transition area between tundra and boreal coniferous forest that ranges in width from 10 to 50 km, maximally 300 km (Fig. 8.9). Trees are sparsely distributed or, more typically, there is a mix of tundra and forest in which the areas of tundra increase towards the pole and of forest southwards (Fig. 8.10).

The *polar tree line* (Fig. 8.2) links the most northerly occurrence of individual or small stands of trees. Usually only trees of at least 5 meters in height are

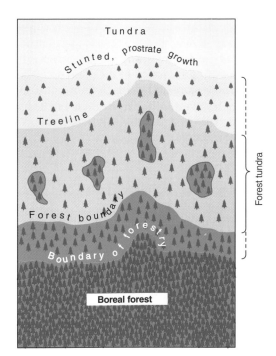

Fig. 8.10. Forest and treeline in the transition area between boreal coniferous forest and tundra. Source: Hustich 1966

included. The *forest line* is the northern boundary of areas of more or less continuous forest cover. Conifers dominate everywhere except in the maritime areas of Scandinavia, Greenland and Kamchatka where there are also stands of birch. Larsen (1981) defines the forest tundra as beginning where the forest cover falls below 75%. Areas that are treeless because of the soil conditions are not included in the estimate.

The location of the polar tree line is determined by the duration and extent of the summer warm period. If the growing season has a mean of $t_{mon} \geq 5\,°C$ over less than 4 months, or a diurnal mean of $\geq 5\,°C$ over less than 105–110 days, or no month with more than $t_{mon} \geq 10\,°C$, or the sum of all diurnal means $> 0\,°C$ is less than 600, trees cannot develop cold new shoots and assimilation organs to the point where they will be able to withstand the cold, dry winter to come. In addition, the ability of trees to produce seeds or of seeds to germinate is limited to summers or a series of summers in which the temperature lies considerably above the mean, or when the minimum temperature for germination, especially in the soil, is exceeded. Towards the tree line the frequency of exceptionally warm summers falls to near zero. Most trees in the forest tundra tend to belong, therefore, to one of the few years in which conditions were suitable for germination. A consequence of this is that stands of similar age are characteristic of the forest tundra.

8.5.4
Biomass and primary production

The *biomass* of a mature boreal coniferous forest ranges from about $150\,t\,ha^{-1}$ in the north of the ecozone to about double this quantity in the south. The increase in height of the trees and densities of the stands from north to south reflects the progressively more favorable climatic conditions for growth.

Primary production of the plant cover is limited everywhere in the zone by climatic factors and the lack of mineral nutrients. The mean annual production of biomass is little more than 4 to $8\,t\,ha^{-1}$, whereby the larger amounts are attained only on the nutrient richer, warmer soils of south facing slopes in the southern taiga.

8.5.5
Decomposition, organic soil matter and mineral reserves

The *decomposition rate* of organic waste is very low, much lower than in a deciduous forest of the midlatitudes (see also Chap. 9.5.4). *Fire* plays an important role in decomposition. The mean reoccurrence interval of forest fires is only 50 to 100 years. Lightening is the most common cause, in contrast to the savanna and tropical rainforest where fires are usually started by man. Forest growth on the burnt over areas benefits from the increased supply of minerals available following burning as well as the lowering of the permafrost table and increased decomposition rates that result from the soil being heated to a higher temperature and greater depth by the fire. In the first stages of

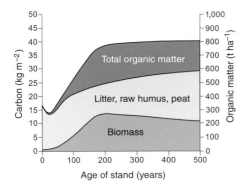

Fig. 8.11. Changes in biomass and soil organic matter in relation to the ages of stands in boreal coniferous forest. Because of its very low rate of decomposition, organic carbon in the soil might increase for several hundred years before there is an equilibrium between delivery of organic matter and decomposition. In Fig. 8.11 this has not ocurred after 500 years and the forest remains a net sink for carbon. The biomass does not increase after 200 years, reaching a level of about 300 t ha^{-1}. It then declines slowly. Source: Kasischke et al. 1995

regeneration after a fire, a cover of shrubs develops in which more demanding broadleaved species such as poplar and birch dominate. There is also a greater variety of fauna.

The greater the amount of dead organic matter in the soil, the longer the period since the last forest fire. After a long absence of forest fires, the layer of litter on the ground in old stands can reach up to 1,000 t ha^{-1}, far exceeding the quantity of living organic matter in the vegetation (Fig. 8.11).

Large amounts of carbon and mineral nutrients are bound in the dead organic matter; with the development of peat, some minerals are even lost from the cycle for ever, particularly the available stocks of nitrogen, calcium and to a lesser extent, potassium. Shortages that result from insufficient nitrogen recycling are partly compensated for by retranslocation from needles and leaves which covers 30% to 40% of nitrogen requirements, also by a very high nitrogen use efficiency of over 200 and by the N$_2$ fixation by blue-green algae which contributes considerably to the supply of anorganic nitrogen (ammonium).

The volume of soil affected by decomposition and the rate of decomposition depend on the depth of the summer thawed layer (active layer) and the extent to which it is heated. Because of the large quantities of humus stored in the Boreal zone soils, it is likely that changes in the climate will have a major effect on the release or binding of carbon, and consequently the CO$_2$ balance in the atmosphere. The *global warming* that is believed to be taking place is, in large part, thought to be caused by the measurable increase in CO$_2$ in the atmosphere (Fig. 8.13). Should this be correct, an increase in the amount of carbon released in northern areas would help to accelerate global warming, which, in turn, could further increase the rate of carbon release.

It is questionable whether such positive feedbacks would develop. If a long lasting warming of the climate occurs, it is probable that the production capacity and area of the boreal forest and tundra would also increase and the forest boundary move closer to the poles. An increase in the amount of organic bound carbon in the biomass would also then follow and the former net release from dead organic soil matter possibly compensated for (Kolchugina and Vinson 1993 and Shaver et al. 1992).

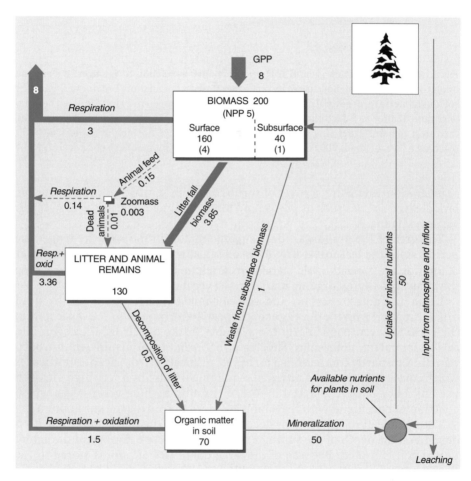

Fig. 8.12. Model of ecosystem of a boreal coniferous forest. Characteristic for these forests is that the large supply of dead organic matter, litter and humus in and on the soil, is approximately equal to the supply of living matter. Two-thirds is litter, which is decomposed at a rate of only 3% per year. There are few animals. Fire is more important for nutrient recycling than animal feeding. The supply of minerals available to plants in the soil is small. Width of arrows, areas of circles and boxes are approximately in proportion to the volumes involved. Organic substances in $t\,ha^{-1}$ or $ha^{-1}\,a^{-1}$, minerals $kg\,ha^{-1}\,a^{-1}$. Source: Persson 1980, Shugart et al. 1992

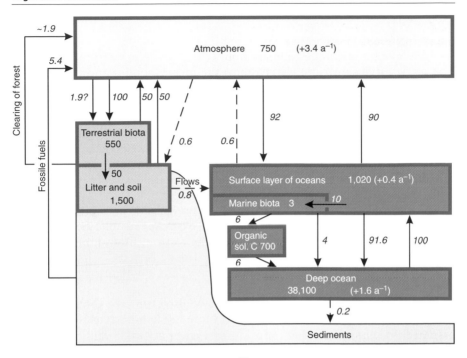

Fig. 8.13. Global carbon circulation. Gt C = 10^{15} gC. Data is for the period 1980–1989. The man-made emissions of carbon dioxide, largely the result of the burning of fossil fuels (5.4 Gt) and the clearing and burning of tropical forests (1.9 Gt) amount to an estimated 7.0 ± 1.1 GtCa^{-1}. About 3.4 Gt is taken up by the atmosphere, in which carbon dioxide is increasing. About 2.0 Gt $\pm$ 0.6 are absorbed by the ocean, mostly the deep oceans. The remaining 1.9 Gt is not fully accounted for. Some may be taken up by the boreal forests and peat bogs and in the tundra. Probably CO_2 released by burning is less than estimated, with some of the amount going into an increasing biomass, the result of the improved CO_2 supply in the primary production, larger supplies of nitrogen oxide from the atmosphere and higher applications of fertilizer. Also, in North America, fire protection measures and reaforestation have resulted in an increase in the wood mass. The missing sink would probably be equal to the CO_2 stored by the boreal and subpolar ecosystems, estimates for which range from 0.5–0.8 Gt. Source: Siegenthaler and Sarmeinto 1993

Because of the greater nitrogen supply from the atmosphere, the growth in biomass will probably be reinforced, if it has not already occurred since man has largely been the cause of the increase by burning, cultivating legumes and fertilizing his fields. The estimated amount of biologically usable nitrogen (nitric oxide, ammonium) induced by man world wide ranges from 132 to 153 million t a^{-1}, compared to the 140 million t a^{-1} that are created naturally (Vitousek et al. 1997). The mean supply of nitrogen to plants has almost doubled in recent decades. Regionally, this has been concentrated in northern and central Europe.

8.5.6
Boreal coniferous forest ecosystems

The model in Fig. 8.12 attempts to show the mean supplies and turnovers of boreal coniferous forest ecosystems in conditions of steady state. The assumptions in the model are, in fact, unrealistic since both the tundra (Fig. 7.18) and the boreal forests and peat bogs are exceptional among ecosystems in that they probably continuously produce a surplus of organic material (Apps and Price 1996, see also Fig. 8.11). All other ecozones show a balance in the medium term. Bird et al. (1996) have confirmed this supposition on the basis of C^{13} and C^{14} measurements. Apps et al. (1993) estimate the annual storage of org. C at 0.7 Gt in the Boreal zone and 0.17Gt in the tundra.

These estimates would indicate that in the Boreal zone and the tundra over 400 Gt (Waelbroeck (1993) or possibly even 700 Gt (Apps et al.(1993) of carbon are bound in the organic soil material in the form of litter, humus or peat. A very large amount when compared to the estimated 1,500 Gt supply of org. C in soil worldwide and the current approximate 750 Gt of anorg. C (as CO_2) in the atmosphere (Fig. 8.13).

Price and Apps (1995) and Waelbroeck have also estimated about 100 Gt for carbon stored in the vegetation (standing biomass) in both ecosystems. If it is assumed that the tundra and boreal forests cover about 25 million km^2 and that 1 g org. C corresponds to a 2.2 g of dry matter in a biomass, a total of 100 t of biomass per hectare results. This value is certainly below the actual value for forests, but when approximately 10 million km^2 of biomass poor tundra are taken into account, the 100 Gt is probably a realistic estimate.

The stored carbon in the soils and biomass of the high latitudes total about one-third of the available supplies on the landmasses of the world although the boreal forests and tundra cover only one-sixth of the total area. On the basis of these data, it would appear that the protection of the boreal forests is of similar importance to the protection of tropical rainforests.

8.6
Land use

Although rich in mineral resources, the Boreal zone coniferous forests are one of the least densely populated regions of the world, with fewer than 5 inhabitants per km^2 in most areas. It is also one of the regions least affected by man. The felling of timber, peat cutting, as well as traditional hunting for furs and the collecting of berries are the main economic activities. Agriculture is unimportant but it is hoped that the management of game and tourism have a future.

The production of timber covers 90% of the world's requirements for paper and saw timber. These volumes must be seen in relation to the vast area covered by forest. The actual area affected by forestry is small. It is estimated that the boreal forests contain three-quarters of the world's reserves of softwoods.

The problems related to commercial forest increase from south to north, included are

1. Distance and consequently long transportation routes to manufacture and consumption centers. Rivers flow northwards and cannot, as elsewhere, be used for timber transport. Few people are available to work in the forests.

2. Low temperatures and deep snow in winter.

3. Low usable quantities of timber per surface area, low quality of timber, stands are not dense, trees are not tall.

4. Low growth capacity. Regeneration of the forest or reafforestation require longer periods than in the midlatitudes before timber can be cut again. In order to retain reserves of timber, it is essential not just to exploit forests, as has usually occurred, but for there to be simultaneous reaforestation.

The working of peat is particularly important in the Eurasian part of the Boreal zone (Paavilainen and Päivänen 1995). In the area of the former Soviet Union reserves are estimated at 200 to 250 billion tons, 66% of the world total (Gore 1983). Worldwide 170 million tons were produced in 1984 and 217 million in 1989. About 90% was used to improve soil for cultivation and the remainder as fuel in power plants or for heating buildings in remote settlements.

Agriculture is possible in permafrost areas if the depth of thaw in summer is at least one meter. The polar limit for cultivation lies about 5° to 10° latitude south of the forest boundary Spring barley is the northernmost cereal crop, maturing in 90 to 95 days, with a limit of cultivation in northern Europe of about 70°N. Spring oats and rye are grown further south. Both are less demanding on the soil and can be grown on nutrient poor Podzols. Potatoes are the most northerly root crop and are also grown up to 70°N in Scandinavia. Most northern forest areas in Eurasia are used as pasture for reindeer. In North America these areas are unusable for most forms of agriculture. Because of the limited agricultural potential of the Boreal zone, other forms of animal husbandry such as a controlled use of the undomesticated caribou and the management of moose and American elk are being examined.

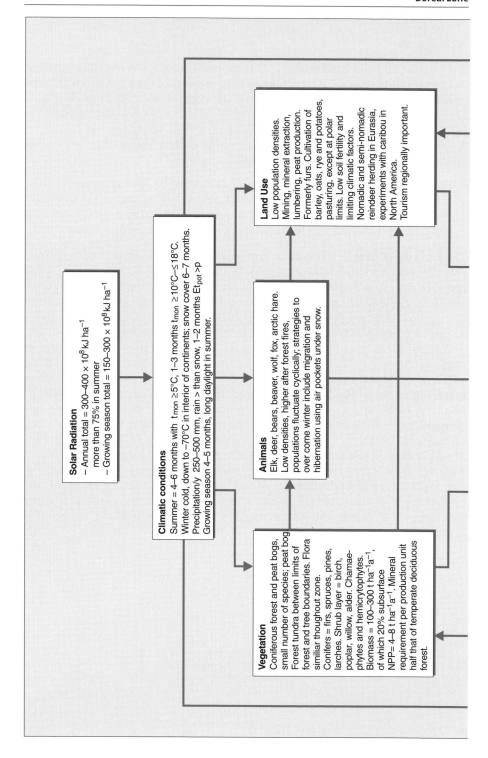

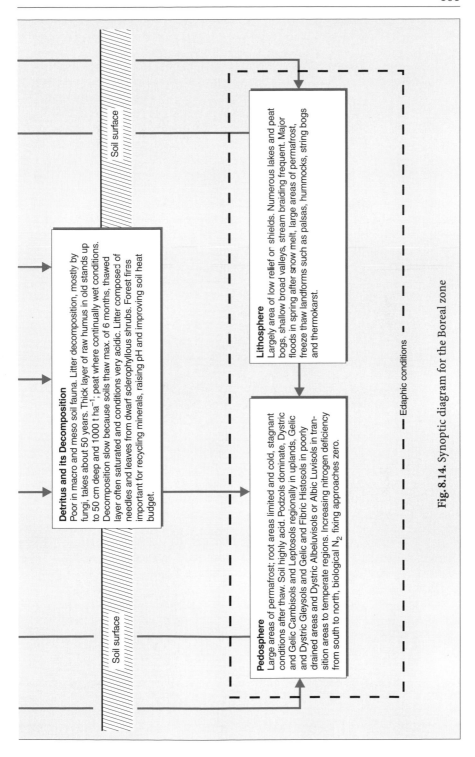

Detritus and its Decomposition
Poor in macro and meso soil fauna. Litter decomposition, mostly by fungi, takes about 50 years. Thick layer of raw humus in old stands up to 50 cm deep and 1000 t ha⁻¹; peat where continually wet conditions. Decomposition slow because soils thaw max. of 6 months, thawed layer often saturated and conditions very acidic. Litter composed of needles and leaves from dwarf sclerophyllous shrubs. Forest fires important for recycling minerals, raising pH and improving soil heat budget.

Lithosphere
Largely area of low relief or shields. Numerous lakes and peat bogs, shallow broad valleys, stream braiding frequent. Major floods in spring after snow melt, large areas of permafrost, freeze thaw landforms such as palsas, hummocks, string bogs and thermokarst.

Pedosphere
Large areas of permafrost; root areas limited and cold, stagnant conditions after thaw. Soil highly acid. Podzols dominate, Dystric and Gelic Cambisols and Leptosols regionally in uplands, Gelic and Dystric Gleysols and Gelic and Fibric Histosols in poorly drained areas and Dystric Albeluvisols or Albic Luvisols in transition areas to temperate regions. Increasing nitrogen deficiency from south to north, biological N₂ fixing approaches zero.

Soil surface

Edaphic conditions

Fig. 8.14. Synoptic diagram for the Boreal zone

CHAPTER 9

Temperate midlatitudes

9.1
Distribution

Most of the Temperate midlatitudes lie in the Northern Hemisphere in eastern and western North America and Eurasia. In the Southern Hemisphere only small areas in South America, Australia and New Zealand fall into the zone. The latitudes in which the zone occurs vary depending on the influence of cold or warm ocean currents, on the west side of the continents it lies between 40° and 60°N or S and on the eastern side between 35° and 50°N or S (Fig. 9.1). The total area of the ecozone is about 14.5 million km², 9.7% of the landmass of the earth. The Boreal zone borders the area towards the poles and the Dry subtropics with winter rain towards the equator on the west of the continents and the Subtropics with summer rain on their eastern side. In the center of the continents, the Temperate midlatitudes ecozone is present, as at most, a narrow strip between the boreal coniferous forests and the cold steppes.

9.2
Climate

The *seasonal change in temperature* is pronounced. However the temperatures do not drop as low as those found in the neighbouring Boreal zone to the north and do not climb as high as those found in the subtropical ecozones to the south. The diurnal temperature ranges – greater than in the Polar/Subpolar and Boreal zones but smaller than in the arid regions of the midlatitudes and tropics/subtropics – also occupy an intermediate position. Viewed from this standpoint, the thermal conditions in the Temperate midlatitudes can be classified as *temperate*, as in the case of the common climatic designation for this zone, namely the cool temperate (forest) climate, or as in the case of vegetation designations, e.g. temperate deciduous forest, which is the prevailing zonal plant formation in the Temperate midlatitudes (Fig. 9.2).

In most continental locations the *growing season* lasts for only half the year with temperatures in winter falling to $-30\,°C$. The growing season in areas near coasts may last for the entire year, with a mean temperature of $> 2\,°C$ in the coldest month, in a few areas $> 5\,°C$, but the summer mean of the warmest month is only $< 15\,°C$ compared to $\geq 18\,°C$ in more continental areas. The

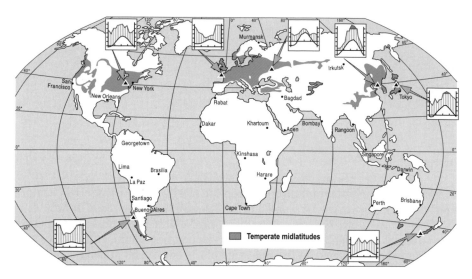

Fig. 9.1. Temperate midlatitude zone

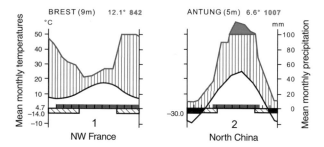

Fig. 9.2. Climates in the Temperate midlatitude zone. 1. Maritime climate in which winters are mild and summers cool. The growing season can be up to 12 months long. Precipitation peaks in winter. 2. Continental climate in which winters are colder and summers hotter than in maritime conditions. The growing season is 6 months. Precipitation usually peaks in summer

range in mean monthly temperatures varies from about 40 °C inland to 10 °C on the coasts where summer temperatures may be insufficient for grain to ripen. In the interior of the continents by contrast, up to three summer months have means of $\geq$ 18 °C. Annual means throughout the zone range from 6 °C to 12 °C.

The length of day varies from less than eight hours in winter to over 16 hours in summer. Approximately 70% of the summer radiation and 50% of the winter radiation reaches the earth's surface directly. Because the sun describes an arc of about 240° in summer, south (or north) facing slopes receive more radiation than north (or south) facing slopes. In Europe, vineyards, for example, are far more common on south facing slopes than those exposed to the north.

Precipitation does not show marked variations seasonally or from year to year. At least 10 months have p (mm) $> 2t_{mon}$ (°C) (Fig. 2.3). Agriculture benefits, therefore, from the rainfall reliability, although in dry periods additional irrigation may be necessary.

Mean annual precipitation ranges between 500 and 1,000 mm in most of the ecozone, only the subtropics and tropics with seasonal or year-round rain have a higher annual precipitaion. A small proportion falls as snow each year. Precipitation totals tend to decline away from the coasts as the depressions move inland.

Figure 9.3 shows the distribution of various parameters in summer within a mixed forest in the Temperate midlatitudes. Most of the radiation is absorbed on the canopy surface (Fig. 9.3A and B) where the highest daytime temperatures occur as well as the largest decrease in humidity, compared to the air above the canopy and below in the area of the trunks. At night the radiation loss from this layer of the plant cover is high, resulting in a reversal of the vertical gradient in the forest.

The amplitudes of both climatic parameters increase with height of the stand (Fig. 9.3C and D). The canopy can be subject to considerable stress from drought or cold, particularly on days with high radiation during the day and major heat loss at night. Within the forest, below the canopy, the climate is more equable, moister and warmer, also the liklihood of frost is less.

Wind does not affect the rhythm of the exchange because the wind turbulence is stopped by the density of the vegetation below the canopy (Fig. 9.3E). At ground level, there is an increase in CO_2 content of the air layer above the soil because of respiration which is reduced again during the day when plants take in CO_2 during the photosynthesis process (Fig. 9.3F). Once the leaves have fallen in winter, the vertical differentiation ceases, or is at most weak.

9.3
Relief and drainage

Both weathering and denudation processes in the ecozone are generally slow so that many of the glacial and glaciofluvial erosion and deposition forms from the Pleistocene Ice Age are still present, as are the remains of Tertiary peneplains.

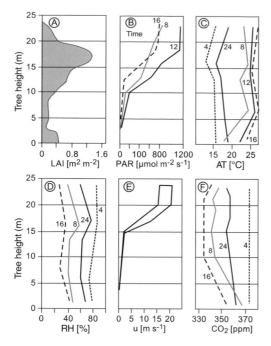

A Layered leaf area index
B Photosynthetic active radiation
C Air temperature
D Relative humidity
E Wind speed
F CO_2 concentration in the air

Fig. 9.3. Vertical profile of climatic parameters in a mixed forest in central Europe in summer. Source: Eliaš et. al

Hydration is important in the weathering of minerals. Bipolar water molecules are deposited on surplus charges of boundary surface cations and the hydration progresses inwards from the outer zone of the rock. The resulting swelling leads to a loosening of the rock structures. *Hydrolysis* can occur simultaneously with hydration. This involves the dissociation of water, whereby H^+ ions can be exchanged with the cations of rock minerals, which also leads to a loosening and eventual destruction of the crystal lattices. The extent of this reaction increases with increasing H^+ concentration, that is, with declining pH value as well as higher temperature. Both hydration and hydrolysis lead to flaking and to the disintegration to grus of the bedrock

Limestone and dolomite are weathered by the process of *carbonation*, the development of carbonate through the solution of carbon dioxide which is present in rainwater and the soil. The CO_2 content is higher in the soil than in the atmosphere because of respiration by soil organisms and roots. The intensity of the solution process increases with the CO_2 content of the water and is promoted by low temperatures. The weathering of limestone bedrock leads to the development of *karst* forms such as dolines and caves with stalactites and stalagmites. In moraines, loess and soils, there is a progressive *decalcification* from the surface downwards.

In contrast to mechanical weathering, chemical weathering in humid climates reaches far down into the soil and bedrock because of the deep percolation of the water. For this reason the soils in the Temperate midlatitudes are more deeply developed than in the boreal forest and polar regions. In the tropical rainforests where temperatures are high throughout the year, the soil layer is often considerably thicker.

About one-third of the precipitation in the Temperate midlatitudes runs off into streams compared to almost 50% in the Boreal zone and over 50% in the tundra. However, because total precipitation is higher in the midlatitudes, the absolute amount of runoff per spatial unit is generally higher (Table 3.1 and Fig. 4.3).

Soil textures are often relatively coarse but the soil structures stable so that the infiltration capacity of seeping water is high. This, together with the closed cover of vegetation means that much of the runoff from the land surface reaches the streams from interflow or the groundwater. It is seldom that the intensity of a precipitation event is sufficient to exceed the maximum rate of infiltration. There is little wash denudation and the peak flow in streams usually occurs several days after a precipitation event.

Stream networks are dense and all streams are perennial, even flowing under ice cover in winter in the interior of the continents. Regimes are less affected by winter frost and spring snow melt than in boreal and polar regions. The year-round precipitation and relatively high evaporation rates in summer play a greater role in the pattern of flow during the year with the result that the summer discharge minimum is more strongly expressed than the frost and snow affected winter minimum. The seasonal peak flows follow precipitation maxima. In the areas of oceanic climate the peaks are in spring and in the interior of the continents, in autumn. The erosive effectiveness of streams is limited and rates of vertical stream erosion are low.

9.4
Soils

The conditions for soil development are more favorable in the Temperate midlatitudes than in all other forest climates. Acidity is low and the decomposition process leads to a higher quality of humus compared to the raw humus of the Boreal zone. In addition, in the cool moist conditions of the zone, three layer clay minerals belonging to the illite and chlorite groups are developed which means that the soils have a much higher cation exchange capacity (CEC) than the poorly absorbing two layer clay minerals in the kaolinte group more common in the tropics and subtropics. Many soils have, therefore, a relatively high natural mineral content which, when combined with fertilizer added during cultivation, can be absorbed and utilized by the crops in greater amounts than in areas with higher mean temperatures.

The most widespread soils in the Temperate midlatitudes are *Haplic* (formerly Orthric) and *Albic Luvisols* (Parabrown earths) and *Eutric Cambisols* (Brown earths). They are often in proximity. Where the surface is pervious and the underlying bedrock has a high $CaCO_3$ content Luvisols tend to predominate, although they are also present on other types of bedrock. Cambisols usually develop on poorer, porous bedrock.

Luvisols are characterized by the translocation of clay (lessivation) from the A to the B horizon with the percolating soil water. This yields the corresponding horizon profile A_h E B_t C. In Haplic Luvisols, the (eluvial) E horizon, whose clay content is depleted, is somewhat lighter brown in color (in the *Albic Luvisols* it is even whitish) than the blackish, humous A_h horizon and the dark brown B_t horizon, in which the illuvial accumulation of clay occurs (= argic B horizon). The Ah and E horizons can together have a thickness of half a meter and the Bt horizon a thickness of up to several meters. Because they are only moderately leached, Luvisols usually have a medium to high base saturation which increases as the climate becomes dryer. The minimum saturation, measured at pH7, is 50% at a depth of 125 cm below the surface in the Bt horizon. The topsoil which is often sandy, and in which leaching has led to a relative increase in coarse grains, can be more acid.

If the E horizon is more strongly bleached, *Albeluvisols* form. They are particularly widespread in the transition areas to the Boreal zone where Eutric Albeluvisols cover extensive areas.

Cambisols have a dark humic A horizon which, with increasing depth, gradually changes over to a B_w horizon that is generally brown in colour (weathered in situ, *cambic B horizon*). This is followed, again without a distinct boundary, by the (unchanged) C horizon. Together the A and B horizons can be up to one and half meters in depth. They generally have a stable structure with favorable water and air budgets. The brown color is due to iron oxides and hydroxides, usually goethite, and is the result of weathering processes relating to iron compounds in the bedrock. The fertile Eutric Cambisols with a base saturation of > 50% develop on basalts and glacial drift. The nutrient poor Dystric Cambisols with < 50% base saturation form on granite and sands. Further development may lead to Luvisols, Podzols and Albeluvisols.

9.5
Vegetation and animals

Forest is the natural vegetation of the Temperate midlatitudes. In the Northern Hemisphere this forest has been almost totally destroyed, either cut for timber, burnt or cleared for agriculture. Only where there is no interest for agricultural or other use, are there forests, mainly by efforestation and for commercial exploitation. At the present time, the total cover of forest compared to earlier periods is small, also compared to all other forest ecozones except the Subtropics with year-round rain. Flora in the Northern Hemisphere belong to the Holarctica region and in the Southern Hemisphere to the Antarctic flora. Flora and fauna in each hemisphere are therefore autonomous and despite their enormous east west extent, quite similar within the ecozone.

In the past and at the present, deciduous or mixed forest of broadleaved trees and conifers predominate. Exceptions are the Pacific Northwest in the U.S.A. where there are pure stands of conifer and parts of the Southern Hemisphere where rainforests of evergreen broadleaved trees have developed, as they had formerly in some coastal areas in western Europe.

9.5.1
Seasonality in deciduous forests

The seasonality related to temperature changes determines the annual cycle of the vegetation in the ecozone. All ecozones in which seasonal changes in moisture supply or temperature occur during the year alter their appearance seasonally. In both the Temperate midlatitudes and the Tropics with summer rain, the changes in appearance of the vegetation during the four seasons is a specific characteristic of the zone. The subtropics and tropics with year-round moisture and the extreme deserts of the dry areas are relatively unaffected as are areas with a high proportion of evergreen species such as the Subtropics with winter rain, the Boreal zone and the tundra.

In spring in the Temperate midlatitudes the vegetation is charactetized by

1. the germination of seeds from the previous year;

2. the production of new shoots, leaves and buds by trees and shrubs;

3. the beginning of shoot growth of hemicryptophytes and geophytes;

4. the emergence of animals from hibernation or dormancy, particularly insects;

5. the return of migratory birds;

6. an increase in birdsong;

7. the mating of most animals and production of young.

Plant growth is initiated on the floor of the forest in spring where light falls through the bare branches of the trees unhindered and warms the leaf litter and upper soil surface more rapidly than the canopy. Numerous spring flowering plants appear on the woodland floor. Some of these plants have intact root systems, hemicryophytes for example, others grow from bulbs, rhizomes and tubers (geophytes) below the soil surface. As leaves and new shoots develop in the shrubs, the radiation absorbing layer moves upwards until the canopy with its new leaf growth is reached.

In summer the trees are in full foliage. Tree trunks and branches thicken, fruits mature and seeds ripen. When the forest canopy is dense, limited light reaches the forest floor, in some cases as little as 10%, so that shade plants with low primary production predominate.

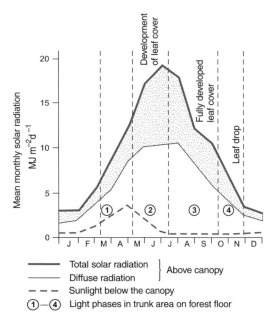

Fig. 9.4. Solar radiation and light intensity during a year in a deciduous woodland in Cambridge, England. Light intensity in the canopy differs from that on the woodland floor. There are four main phases of light intensity: 1. spring phase while trees are bare of leaves; 2. transition phase of buds opening on trees coming into leaf; 3. summer shade phase; 4. autumn phase during leaf fall. There are no light phases on the floor of a coniferous forest. Source: Larcher 1994, Walter and Breckle 1983

In autumn most of the summer production of leaves, seeds and fruits of the woody plants and stems of the herbaceous ground cover falls to the forest floor. Before leaf fall there is a decline in organic matter and a partial resorption of nitrogen, iron, phosphorus and potassium into the branches and trunks. The resulting change from green to pale yellow in alders, ash or willow, for example, or red and gold in maple and birch give the woods their colorful autumn aspect. Insect eating migratory birds leave already in late summer or early autumn. Other bird species from higher latitudes migrate through the region or remain to winter.

By winter, photosynthesis ceases and biological and chemical processes in the soil slow down. Animals such as badgers and squirrels become dormant, others, including dormice and hedgehogs, hibernate. Amphibians, snakes, insects and spiders living on the surface hibernate or leave eggs, larvae or chrysalis which winter in the forest floor or in the barks of trees. Non migratory birds and visitors develop thick plumage and deer, wild boar, fox, weasels and other active animals produce a thicker pelt.

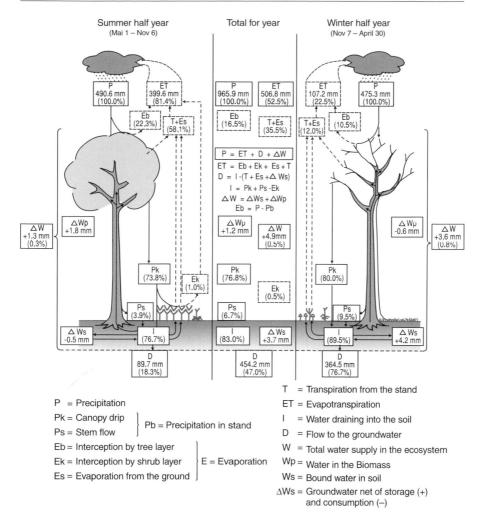

Fig. 9.5. Water budget in a Belgian oak forest, with and without leaf cover. Mean annual precipitation 1964–1968 = 965.9 mm of which 506.8 mm (52.5%) evaporated, 454.2 mm (47%) went into runoff and 4.9 mm (0.5%) was stored in the biomass. Precipitation totals are almost the same in summer and winter. Losses from interception and transpiration are four times higher in the summer half year than in winter: 399.6 mm compared to 107.2 mm. More water drains into the soil in winter than in summer. Because there is a litter layer year-round and shadow in summer, evaporation from the surface is low. Source: after Schnock 1971

9.5.2
Water budget in forests

The water budget in deciduous forests is also subject to seasonal change. Winter surplus increases the reserves and supplies the groundwater. In summer consumption can lead to deficits (Fig. 9.5). The spring water budget is usually in equilibrium in most years with the soil remaining saturated during this period. The evergreen coniferous forest water budget differs in that the stem flow of pines is much less than in deciduous woodland and near zero for spruce because the branches divert water from the trunks. Also interception at the crown of the spruce is greater in summer and winter because it is evergreen, than the deciduous species. It has been estimated that annually 27.2% of the precipitation is intercepted by spruce compared to 17.1% for beech (Fig. 9.5).

Transpiration losses by spruce and beech also differed in relation to the plant matter produced during the period studied. Spruce required 220 liters per kg of dry matter compared to 180 liters for beech, which means that the latter has a much higher water use efficiency and lower transpiration coefficient. By comparison, ground cover plants require 300 to 400 liters per kg of dry matter (Ellenberg et al. 1986).

9.5.3
Biomass and primary production, growth and litter production

The *biomass* increases over many years as the age of the stand increases. Depending on the life span of the tree species in the stand, the maximum biomass is reached after 100 to 200 years. In most cases the total lies between 200 and 400 t ha^{-1} of which 20% is root matter. In Fig. 9.6, primary production peaks after 40–55 years with about 11 t ha^{-1} a^{-1}. During this period the number of species of vascular plants is greatest because species from earlier stages in the development of the forest are still present.

The division of the *primary production* into growth and litter production is also dependent on age (Chap. 5, Fig. 5.3). Initially growth clearly dominates.

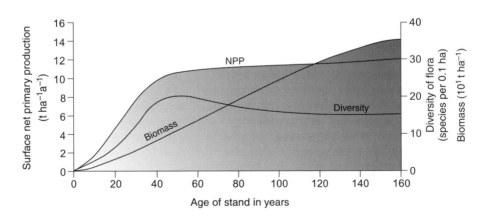

Fig. 9.6. Changes in primary production, biomass and species diversity in an oak–pine forest in the eastern U.S.A. in relation to the age of the stand. Source: Whittaker 1970

Later waste increases in relative importance but until the stand is old is it less than the contribution from growth in terms of quantities. As long as trees do not fall and only fine litter accumulates, leaves supply, with a production of 2 to $4\,t\,ha^{-1}\,a^{-1}$, between 60 and 80% of the litter. Scale from buds, blossoms, fruits, bark and twigs account for the remainder.

9.5.4
Mineral budget in midlatitude deciduous broadleaf and boreal coniferous forests

The data on *mineral stocks* and *turnover* used in this section and in Fig. 9.7 are based on values from 14 deciduous forests in Europe and North America (International Biological Programme, Cole and Rapp, 1981, and De Angelis et al. 1981). Cole and Rapp also provide data from three areas of boreal forest in Alaska in which spruce dominates.

The mean mineral content in the tree layer in both boreal coniferous forests and deciduous forests is 1%, although the latter has a higher absolute quantity of mineral reserves because the biomass is greater. Calcium is the most frequent nutrient element in both types of forest followed in declining amounts by nitrogen, potassium, magnesium and phosphorus.

Although the mean is 1% for the whole forest, there is a considerable range in the content of different minerals within its components parts. In general, the leaves have a higher mean content than bark or wood. Figure 9.7 shows deciduous foliage with a mineral content of 4.3% in dry matter compared to 0.6% for trunks.

With the exception of calcium, mineral concentrations decline during the summer due to losses from leaching and translocation, although the sequence of concentrations remains the same until after the leaf drop in autumn.

The *net primary production* of the tree layer in deciduous forests has a mean of $10\,t$ dry matter $ha^{-1}\,a^{-1}$. Of this 40% is involved in the production of foliage seasonally and not related to long term growth of the forest cover. Because of the high mineral content of the leaves, they use 80% of the total minerals required for primary production in the tree layer.

In the coniferous forests where the turnover rate of needles is much lower, mineral poor wood is produced. Only one quarter of the tree layer net primary production and 50% of the required minerals are used in needle production.

With the exception of phosphorus, the net primary production of the temperate deciduous trees requires considerably more minerals per production unit than the boreal needle trees. The deciduous forest produces a mean of only 103 kg of organic matter per 1 kg of nitrogen and the boreal forest double this amount. Boreal forests have, therefore, a much higher mineral use efficiency and particularly a higher nitrogen use efficiency. The high nitrogen use efficiency does, however, produce nitrogen poor plant tissue that decomposes only slowly after the plant dies, which in turn slows down the nitrogen circulation in the ecosystem.

Since the deciduous forests have a higher net primary production, the differences in mineral requirements per spatial unit are even greater. Assuming a net

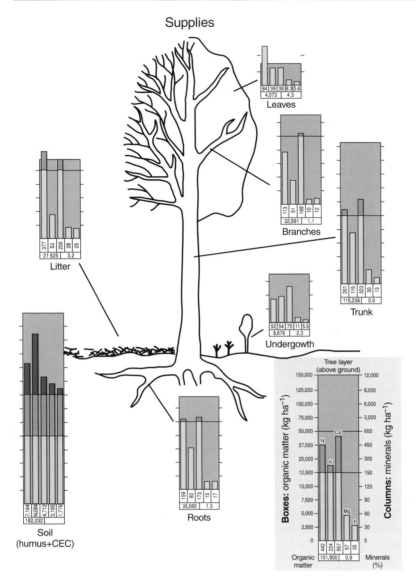

Fig. 9.7a. Supply of organic matter and minerals in temperate midlatitude deciduous forests. The organic matter content and the minerals are shown for the components of the forest. Mean values based on 14 individual stands in Europe and North America. The scales are reduced from 12,500 kg organic matter and 150 kg minerals by 1:5 and from 50,000 kg organic matter and 600 kg minerals by 1:100. The available supply of the tree layer is not equal to the sum of the individual supplies because there were only 12 values for branches and trunks. The values for minerals for the humus and CEC are taken from the soil root zone. Source: Cole and Rapp 1981

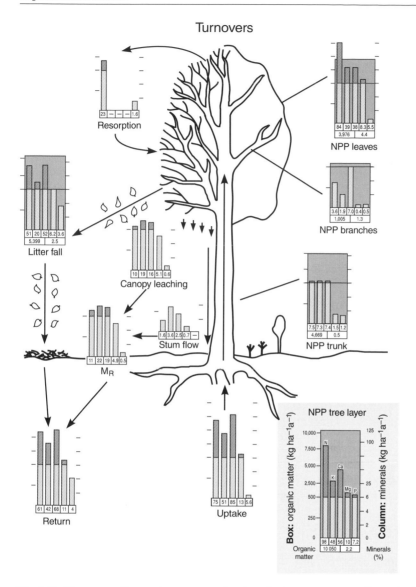

Fig. 9.7b. Turnover of organic matter and minerals in temperate midlatitude deciduous forest. The organic matter and mineral turnovers are shown for the components of the forest. Mean values based on 14 individual stands in Europe and North America. The scale is reduced from 500 kg organic matter and 6 kg minerals to 1:12.5. Differences between the values for the tree layer NPP and the return are due to lack of data. Source: Cole and Rapp 1981

primary production twice that of the boreal forests, the mineral requirements for deciduous cover are four times greater per spatial unit. Consequently, in

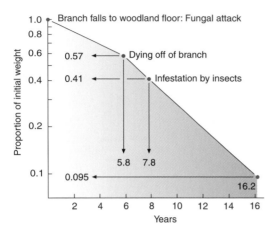

Fig. 9.8. Rate of decomposition of branch wood of more than 2 cm diameter in a temperate midlatitude woodland of oak, ash and birch and hazel in England. The decomposition lasted 16 years. In the 6 years between the dying of a branch and its falling to the ground, 40% of the wood is decomposed by fungi, an annual rate of loss of 8.4%. On the woodland floor, the decomposition rate is 17.1% annually, in part, because of the greater activity of wood boring insects. Source: Swift et al. 1976

deciduous forests the level of plant nutrient supply in the soil is of necessity higher. An annual primary production of 8 to 12 t utilizes 80 to 120 kg nitrogen of which 60 to 90 kg is from the soil and the remainder from the leaves before they fall in autumn. This compares to 20 to 40 kg nitrogen for boreal forests with a production of 4 to 8 t and is one reason why boreal conifers can grow in marginal locations.

In deciduous forests the mineral rich foliage forms a large proportion of the annual accumulation of litter. The proportion of minerals returned is therefore also high. In the 14 woodlands examined (Fig. 9.7), the total mean litter fall including the woody components amounted to $5.4 \, \mathrm{t \, ha^{-1} \, a^{-1}}$ with a mean mineral content of 2.5%. This results in an annual mineral return per hectare of 135 kg, at least 10% of the above ground pools.

In addition to the organic detritus, significant quantities of minerals are leached from the canopy, including up to 20% of nitrogen and phosphorus, 30% of calcium, 40% of magnesium and 60% of potassium. The total amount is about 50 kg or about one quarter of the mineral return. Annually about 80% of the nutrients are returned, although the litter fall during the period forms only 54% of the net primary production of the tree layer.

The litter fall in a deciduous forest forms a closed cover, in Fig. 9.7, this amounted to $21.6 \, \mathrm{t \, ha^{-1}}$. The litter layer is generally only a few centimeters thick because the accumulating litter is usually decomposed rapidly. In boreal coniferous forests the delivery of litter is much smaller but the depth of the litter layer much greater because of the low decomposition rates. Mineral content is only 1% in the needle litter compared to 3.2% in broadleaf litter.

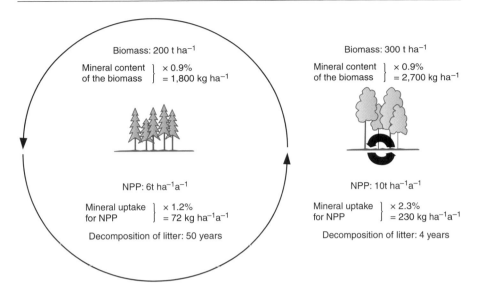

Biomass: 200 t ha^{-1}

Mineral content × 0.9%
of the biomass = 1,800 kg ha^{-1}

NPP: 6t ha^{-1}a^{-1}

Mineral uptake × 1.2%
for NPP = 72 kg ha^{-1}a^{-1}

Decomposition of litter: 50 years

Biomass: 300 t ha^{-1}

Mineral content × 0.9%
of the biomass = 2,700 kg ha^{-1}

NPP: 10t ha^{-1}a^{-1}

Mineral uptake × 2.3%
for NPP = 230 kg ha^{-1}a^{-1}

Decomposition of litter: 4 years

Fig. 9.9. Mineral circulation in a deciduous forest in the Temperate midlatitudes and coniferous forest in the Boreal zone. The % mineral content in the biomass in both types of forest is similiar but larger in quantity in the deciduous forest because of its larger biomass. Uptake, requirement and return of minerals are also much higher and the decomposition of the litter much more rapid than in the coniferous forest. In the figure, the quantity of NPP is assumed to be equal to the waste production and the mineral uptake equal to the return of minerals

If a dynamic equilibrium is assumed in which the same quantity of litter is decomposed as accumulates, the mean turnover duration in a deciduous forest is four years and in a coniferous forest 350 years. There can be wide variations from these means, depending, for example, on the proportion of dead components, the plant composition and the distribution on the soil surface or at depth. In addition, the decomposition of coarse litter takes longer than fine litter (Fig. 9.8).

Leaves from deciduous trees are decomposed within 18 months to 3 years. In central Europe, alder decomposes most rapidly followed by elm, hornbeam, linden, maple, ash, birch and, the slowest, oak. Fungi and bacteria are the most important agents in the decomposition of organic matter in the litter layer and A horizon. Their contribution to the respiratory carbon dioxide of the living soil organisms (edaphon) is 90%. Their share of the biomass of the edaphon is only slightly less. Half the remaining 10% is accounted for by earthworms and the other half by all other soil organisms.

In summary, the deciduous forests of the Temperate midlatitudes have a short but relatively rapid mineral circulation. There is a high rate of uptake of mineral nutrients in spring and summer, most of which returns to the ground in the autumn with the leaf fall and is released again after a mean period of four years from the litter. The litter may include woody components.

In the coniferous forests of the Boreal zone the circulation of minerals takes place over a longer period and at a low rate. The requirement for the net primary production is low and relative to wood, the mineral rich needles last for many years so that the annual loss of minerals is small. The release of minerals from the organic matter does, however, take far longer. Bottlenecks in the nutrient supply occur, therefore, more often in boreal regions than in midlatitude forests.

9.5.5
Model of an ecosystem of a deciduous forest

Figure 9.10 shows a model of the available supply and turnover in an ecosystem for a deciduous forest in the Temperate midlatitudes in a steady state.

Compared to the boreal forest, the litter layer is much less and the deep layer of raw humus typical for the boreal forest is absent. The humus content of the soil is much higher and of better quality. The biomass comprises considerably more than half the total organic matter of the ecosystem. In the boreal forest dead organic matter dominates.

9.6
Land use

The Temperate midlatitudes are more densely populated than in relation to their overall surface. Of the largest most densely populated centers of mankind i.e. (1) Central Europe, (2) East Coast of USA, and (3) South East and East Asia, the two first named areas lie completely within its distribution boundaries and the third with larger parts (of Japan, Korea and China). Changes affecting the original natural conditions in the ecozone have, therefore, been far more widespread and far reaching than in other ecozones. Moorlands and valley bottoms have been drained and converted to pasture or arable land. The remaining areas of forest are usually confined to areas of poor soil or steep slopes unsuitable for agriculture or other uses. The straight boundaries separating fields and woodlands are a visual characteristic of the ecozone's landscape

The Temperate midlatitudes are also the most developed regions economically in the world, reflected in living standards, the extent of urbanization, the level of service industries and the degree of participation in world trade.

The reverse side of high living standards is the high consumption of energy and raw materials and the amount of waste produced relative to the population size. But awareness of the growing problems and dangers has increased, and environmental protection and sustainable use of land and raw materials have become major targets in politics and of several non-state organizations. Strategies to deal with the problems have concentrated on recycling waste products, reducing unit energy use in relation to output and reducing of CO_2 emissions.

Agricultural land use in the Temperate midlatitudes benefits from the favorable temperatures and reliability of precipitation during the growing season.

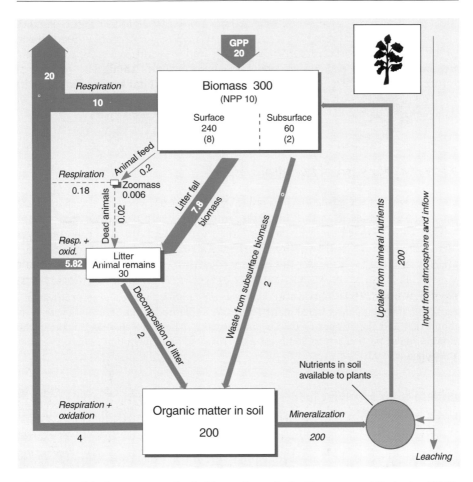

Fig. 9.10. Model of ecosystem of a deciduous forest in the Temperate midlatitudes. Width of arrows, areas of circles and boxes are approximately in proportion to the volumes involved. Organic substances in t ha^{-1} a^{-1}, minerals in kg ha^{-1} a^{-1}. Source: Duvigneaud 1971, Ellenberg et al. 1986, Jakucs 1985, Reichle 1970

Soils are naturally fertile or, with the addition of fertilizer, can be greatly increased in their productivity. The natural potential for agricultural use in the ecozone is high, as is the proportion of land actually used for agriculture (Fig. 9.11).

Mixed farms and intensive animal husbandry are the most common forms of agriculture.

The mixed farms are usually small or middle-sized enterprises that are labor and capital intensive and highly productive. Most frequently grains, root crops and feed crops are cultivated in combination with livestock keeping. Crop production and animal husbandry are closely integrated with, for example,

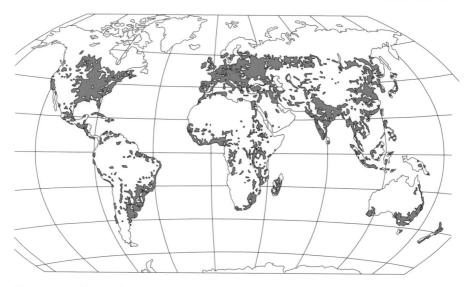

Fig. 9.11. Worldwide distribution of agricultural land (arable, pasture and tree cultivation). Agricultural land is concentrated in the Temperate midlatitudes, the Subtropics with year-round rain and the steppes of the Dry midlatitudes. There are also large areas in the tropical zones, particularly in Southeast Asia (irrigated rice cultivation). Source: Cramer and Solomon 1993

the feed crops supplying fodder for the animals on the farm. A large variety of crops is grown and animals kept. Grains include wheat, rye, barley, oats and in recent decades also maize. Other field crops are potatoes, field vegetables, sugar beet, fodder beet, and oil seed rape. Clover and lucerne are grown for feed. Compared to the Boreal zone, there are more tree crops, although fewer than in the neighboring Subtropics with winter rain or year-round rainfall. Fruits include apples, cherries, pears and plums together with berries such as strawberries and raspberries. Vineyards occur in the warmer areas.

Changes in recent years have lead to consolidation into large units and specialization, similar to the specialized large farms in the Subtropics with year-round rain and Tropics with summer rainfall.

In coastal and upland areas where cool moist conditions favor the growth of grasses, livestock farming with dairy and beef cattle predominates and in some areas also sheep. Pastures are improved with the application of fertilizer and the sowing of feed grasses and clover, which together with drainage measures, increase yields so much that carrying capacities can be increased by as much as 2 to 3 livestock units (1 LU = 500 kg cattle or 5 sheep or goats) per hectare. Intensive pasturing has also developed outside of the traditional areas of grazing using silage crops. Maize cultivation and dairy cattle in the U.S.A. is an example.

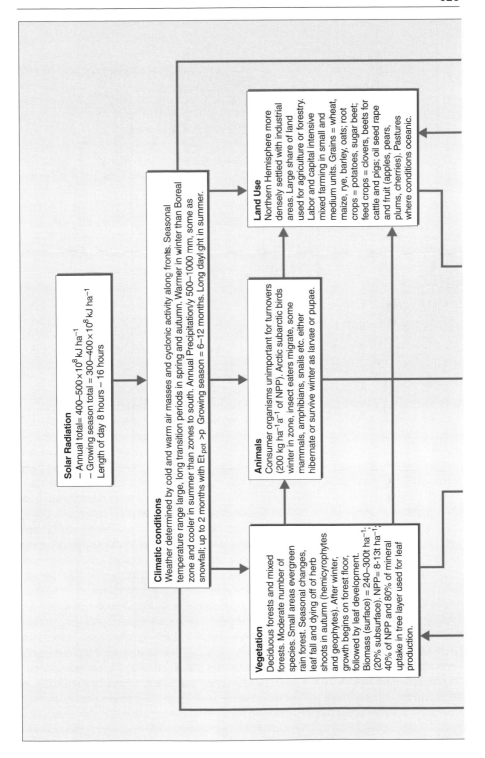

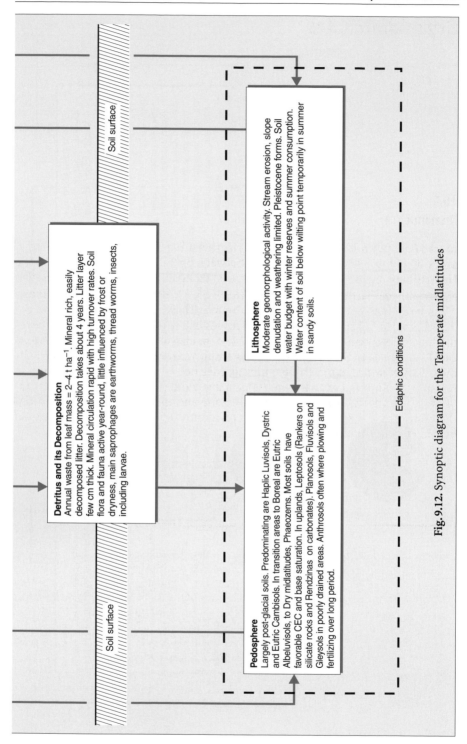

Detritus and its Decomposition
Annual waste from leaf mass = 2–4 t ha⁻¹. Mineral rich, easily decomposed litter. Decomposition takes about 4 years. Litter layer few cm thick. Mineral circulation rapid with high turnover rates. Soil flora and fauna active year-round, little influenced by frost or dryness, main saprophages are earthworms, thread worms, insects, including larvae.

Lithosphere
Moderate geomorphological activity. Stream erosion, slope denudation and weathering limited. Pleistocene forms. Soil water budget with winter reserves and summer consumption. Water content of soil below wilting point temporarily in summer in sandy soils.

Pedosphere
Largely post-glacial soils. Predominating are Haplic Luvisols, Dystric and Eutric Cambisols. In transition areas to Boreal are Eutric Albeluvisols, to Dry midlatitudes, Phaeozems. Most soils have favorable CEC and base saturation. In uplands, Leptosols (Rankers on silicate rocks and Rendzinas on carbonates). Planosols, Fluvisols and Gleysols in poorly drained areas. Anthrosols often where plowing and fertilizing over long period.

Soil surface

Soil surface

— ▪ — Edaphic conditions

Fig. 9.12. Synoptic diagram for the Temperate midlatitudes

CHAPTER 10

Dry midlatitudes

10.1
Distribution

Most of the Dry midlatitudes lie in the Northern Hemisphere in the continental areas of Eurasia and North America between 35° and 50°N. In the Southern Hemisphere, the zone is present only in east Patagonia in South America and in New Zealand. The total area is about 16.5 km², 11.1% of the world's landmass.

Its boundaries can be defined by climate. Thermal criteria determine the boundaries to the Dry tropics and subtropics in Eurasia, in west central Asia, in the U.S.A., between the Midwest and Mexico, and in South America, between east Patagonia and the Pampas. The Dry tropics and subtropics begin where the mean temperature in the coldest month does not fall below 5 °C (therefore no restrictions for plant growth) and at least five months have mean temperatures of more than 18 °C. Towards the Boreal zone and the Temperate midlatitudes the boundary lies where more than four months of the growing season have humid climatic conditions.

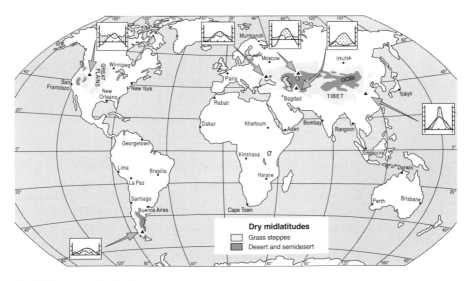

Fig. 10.1. Dry midlatitudes

Within the continents the Dry midlatitudes can be subdivided according to the degree of aridity, the associated plant formations and the possibilities for cultivation. If more than 100 mm precipitation fall in the growing season and 2 to 4 months have some rainfall, there are, or were, steppes on which wheat can be cultivated. Where between 50 and 100 mm rain fall during the growing season, semi-desert predominates and below 50 mm there is desert. In the dry regions of Eurasia, rainfall declines from north to south and in North America from east to west. *Steppes* cover originally three-quarters of the ecozone.

In all dry regions, whether in the midlatitudes, subtropics or tropics, the following features are present:

1. Plant growth is limited to, at most, 5 months of the year because of aridity.

2. There is often a lack of moisture even in the rainy season because of variability in precipitation and also low reserves of water in the soil.

3. Cultivation that relies only on the supply from rainfall brings risk. Alternatives are to use dry farming methods, rapidly growing or drought resistant plants or irrigation.

4. The natural vegetation is composed of xerophytes and halophytes, usually widely dispersed.

5. The net primary production is low. Above ground NPP has been estimated at $3\,t\,ha^{-1}\,a^{-1}$ for deserts, semi-deserts and desert steppes and $6\,t\,ha^{-1}\,a^{-1}$ in the semi-arid transition zones. (Smith and Noble 1986)

6. Streams are ephemeral and generally endorheic

7. The upward movement of water in the soil during dry periods results in an increase in calcium carbonate, sometimes also calcium sulphate, and easily soluble salts in the soil profile. Soils are, therefore, alkaline with high pH values and base saturation of 100%. Salt crusts may also form.

10.2
Climate

The Dry midlatitudes lie in the west wind belt, or cyclonic west wind drift beyond the tropics. The zone is largely continental in location and has longer periods of sunshine, higher solar radiation, lower precipitation and a greater temperature range than the Temperate midlatitudes which also lie in this wind belt.

Precipitation is less than the potential evaporation in most months. Rainfall is highly variable and unreliable at all times of the year. Long dry spells are frequent during the rainy season and deviations from the annual means are considerable.

In addition to the stress caused by drought, is the stress caused by low temperatures. Mean temperatures fall below 0 °C for at least one month of the year and snow cover lasts from a few days to several months.

Solar radiation in midsummer is similar to that in the Dry tropics and subtropics since the longer hours of daylight compensate for the lower angle of the sun. With the exception of Patagonia and New Zealand, the summers are hot. The mean monthly temperature exceeds 20 °C in up to three months of the year in many areas, with monthly maxima up to 30 °C.

10.3
Relief and drainage

Geomorphological processes in all dry regions of the world are similar. Differences are related to variations in the degree of aridity and rock type rather than the temperature differences related to latitude (Chap. 13). Processes and forms that occur primarily in the Dry midlatitudes and not in the other dry areas are described in this Chapter, they include frost shattering, gelifluction and needle ice.

Frost shattering on bare rock plays an important role if there is sufficient water present in the rock. The process is more effective than in the Temperate midlatitudes because freeze thaw is more frequent and the temperature falls further below the freezing point, also, because there is no insulating vegetation cover, the frost reaches deeper into the soil and rock.

Gelifluction takes place during the spring melt when the soil above the still frozen subsurface is saturated. Fields in the steppes are especially liable to this typo of downslope movement. It can be reduced if winter wheat is cultivated.

Needle ice forms if the soil temperature sinks below 0 °C in the uppermost pores of the soil on unvegetated ground following sublimation of the moisture in the cooled air. The needle-shaped crystals grow upward and lift soil particles at right angles to the surface. The soil particles are loosened and may be removed by wash denudation and, in particular, deflation, although biogenic crusts of blue-green algae, lichen and fungus can slow these denudation processes.

Episodic flow resulting from the spring melt of winter snow cover has more influence on drainage patterns and regimes in the ecozone than the summer precipitation.

10.4
Soils in the steppes

10.4.1
Types of soil in the zone

There is little leaching in the soils of the Dry midlatitudes. In general, the *capillary rise* of moisture in the soil that results in deposition of soluble salts, carbonates and gypsum in the upper layers is more important than leaching. Pedocals which are characterized by free calcium in the form of concretions in the profile and a high base saturation are the most frequently occurring soils. The sequence of horizons is Ah-ACk-C or a simplified AC horizon.

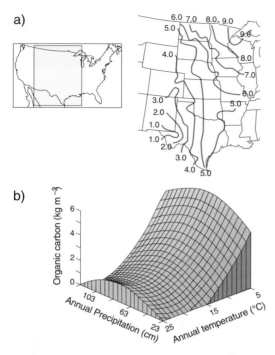

Fig. 10.2. Soil organic matter content (kg m^{-2}) in the upper 20 cm of a loam soil (20% clay, 40% silt) in the Great Plains, U.S.A. in relation to a) regional differentiation, b) temperature and precipitation. Humus content increases eastwards with increasing precipitation and northwards with decreasing temperature. The highest humus content is therefore in the northeast

The amounts of organic soil matter are high. Under otherwise equal conditions, they increase with increasing precipitation and declining mean annual temperature because of the greater primary production and lower rate of decomposition (Fig. 10.2). The type of humus is generally a mull in which highly polymerized humins and interim products of humification, which together with clay minerals form stable nitrogen rich organic mineral complexes (mullic A horizon).

The amounts of dead soil organic matter are high. Under otherwise equal conditions, they increase with increase in precipitation and declining mean annual temperatures because of the greater primary production and lower rate of decomposition (Fig. 10.2): The type of humus is generally a mull in which highly polymerized humins and interim products of humification together with clay minerals form stable nitrogen rieh organic-mineral complexes (mollic A horizon).

The *soil structure* is crumbly and the exchange capacity and water capacity are high, these together form the basis for the *fertility* of these soils. Only aridity limits plant growth. Regionally the sequence of Phaeozems, Chernozems and Kastanozems reflect decreasing humidity in the zone.

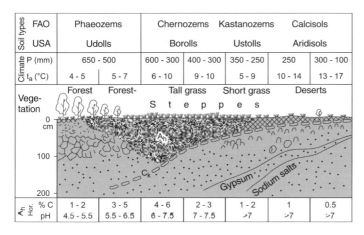

Soil types	FAO	Phaeozems		Chernozems	Kastanozems	Calcisols	
	USA	Udolls		Borolls	Ustolls	Aridisols	
Climate	P (mm)	650 - 500		600 - 300	400 - 300	350 - 250	250	300 - 100
	t_a (°C)	4 - 5	5 - 7	6 - 10	9 - 10	5 - 9	10 - 14	13 - 17

Ah Hor.	% C	1 - 2	3 - 5	4 - 6	2 - 3	1 - 2	1	0.5
	pH	4.5 - 5.5	5.5 - 6.5	6 - 7.5	7 - 7.5	>7	>7	>7

Fig. 10.3. Steppe soils in Russia. With increasing aridity the thickness and humus content (%C) of the Ah horizons increase initially and then decline; also leaching decreases and calcium carbonate (Ck), gypsum, sodium salts and pH values increase. Source Schachtschabel et al. 1998

Phaeozems occur in the steppe areas with the highest precipitation, between 500 and 700 mm annually. The soils are deep, rich in humus, dark brown to grey black in color and formed on base rich sediments, often loess. Decalcification and iron oxidation are well developed and unlike other steppe soils, they have no calcium rich horizon.

Chernozems form in areas where the precipitation ranges from 400 to 550 mm. They are dark in color and have a 50 to 100 cm thick Ah horizon in which the humus content can exceed 10%, although in central Europe the humus content of black earths ranges from only 2% to 6%. The generally favorable structure and high exchange capacity are the consequence of the high humus content. The deep Ah horizon is due to the parent material of the soil, often loess, which has a high calcium carbonate content, the semi-arid and winter cold continental climate, the natural vegetation of grasses and herbaceous plants, and the bioturbation by fauna in the soil.

During the moist warm spring season there is a large production of biomass with a high carbon nitrogen ratio easily used by consumers and detritus feeders. There are, however, also long periods in the hot and dry summer season and the cold winter season during which microbiological activity is delayed. As a result a large proportion of organic matter produced, is accumulated in the humus in which animals that live underground, such as hamsters, ground squirrels and prairie dogs, together with earthworms are active in the mixing process.

Kastanozems replace Chernozems where mean annual precipitation totals between 200 and 400 mm. The Ah horizon is not as deep as in the Chernozems and brown in color (Hence the name "chestnut soil" in the USA). Secondary calcium carbonate and gypsum enrichment occurs at higher levels in the soil than in Chernozems. Because of the aridity, soluble salts are also present in the subsoil. The natural vegetation is a short grass steppe.

10.4.2
Halomorphic soils

Halomorphic soils develop in dry areas in locations that are stagnant or in areas in which the groundwater table is high. In similar locations in wetter climates Gleysols, Planosols, Fluvisols or Histosols are the most common types of soils. Halomorphic soils are characterized by a content of soluble salts or sodium that is high enough to impair plant growth, especially of plants for cultivation.

Solonchaks have a high content of easily soluble salts in the upper Az horizon or the Bz horizon below. The salts are mostly chlorides, sulphates or bicarbonate of sodium and, more seldom, chlorides or sulphates of magnesium and calcium. The properties and characteristics of Solonchaks vary with the type, quantity and distribution of these salts.

Plants can grow on Solonchaks only after the salts have been leached out. With irrigation, new salts are often added. Successful utilization using fresh water is possible only where the water can percolate down through the soil and where the groundwater table is deep (Chap. 13).

Solonetz have a high sodium saturation of the sorption complex, > 15%, in the upper 40 cm of the argic Btn horizon. Solonetz are usually in an area of former Solonchaks which had a high sodium content but have been desalinated following a lowering of the groundwater table or a change to wetter climatic conditions. They are of little use in cultivation because of the strong alkaline reaction and related swelling and shrinking processes within the soil which cause either poor aeration and stagnation or the formation of hard blocks with fissures. The availability of nutrients is also low and the high sodium concentrations have a toxic effect on most cultivated plants. If gypsum is added an exchange of sodium ions with calcium ions is possible, followed by a leaching of the newly formed easily soluble sodium sulphates.

10.5
Vegetation and animals in the steppes

The central areas of the Dry midlatitudes are semi-deserts and deserts They are surrounded by a broad transition area towards the borders of the zone in which precipitation is higher and the vegetation is an open shrubland with a grassy undergrowth, the *shrub and thorn steppes*, or semi-arid grasslands, the *grass steppes* or *prairies* in North America.

The steppes of the Dry midlatitudes have almost no trees, the result primarily of the dominant soil types in these regions which have an exceptionally high water storage capacity, much of it available and of benefit to the densely rooted steppe grasses rather than the deeper and more extensively rooted woody plants. By contrast, in the Dry tropics and subtropics the soil formation processes do not result in a humus rich soil so that the dense root systems of the grasses are much less favorable. Here, under natural conditions, a mix of trees and grasses dominates.

10.5.1
Types of steppes

Depending on the degree of aridity in the ecozone, five types of steppe can be distinguished. *Forest steppe* is present in Eurasia and is a transition zone towards the Boreal zone and the Temperate midlatitudes. It is characterized by stands of dispersed trees and islands of grassland. With increasing dryness, trees are fewer and eventually only islands of trees remain on the grassland. Soils are largely Phaeozems.

Tall grass steppe, also described as moist, herbaceous or meadow steppe, forms a continuous cover of grasses that are at least 50 cm, even up to 200 cm, in height when mature. Islands of woodland are present but limited to stony environments or hollows with a water supply. There are numerous varieties of herbaceous plants, particularly composites and legumes. The leaf area index (LAI) of the herbaceous layer usually has a value of more than 1 and a maximum that is considerably higher, almost double that of the short grass steppe. More than three months in summer are arid but the majority of months has precipitation or snowfall or a precipitation of $> 50\% \, ET_{pot}$. The annual balance of precipitation minus potential evapotranspiration $(P - ET_{pot})$ is at most just negative. Snow melt leads to a thorough wetting of the soil in spring so that there is no lack of moisture at this time of year. The net primary production has a limited nitrogen content. Chernozems are the most widespread soil.

Mixed grass steppes occur in North America in the transition area between moist and dry steppe and are characterized by a mix of grass varieties of middle height and the low grasses of the short grass steppes.

Short grass steppe (or dry steppe) is largely without areas of woodland, apart from some trees along the banks of rivers. Most grasses grow in tussocks and reach 20 to 40 cm in height. The bases of the tussocks cover only about 50% of the surface. The leaf area index is less than 1 and near zero outside of the growing season. Seven to ten months are arid or at least semi-arid and plant growth occurs only in spring. Kastanozems are the predominating soils.

Desert steppes are related in part to climatic conditions but are also the consequence of over grazing. Vegetation consists of dwarf shrubs and semi-ligneous shrubs. Grass growth is sparse. The proportion of annual grasses in relation to perennial grasses is higher than in areas with more precipitation. Plant cover is incomplete with a coverage of $> 50\%$. In general only one month has any rainfall and Xerosols dominate.

The desert steppes are usually included with the semi-deserts or deserts because of the spacious plant stand and the large proportion of woody plants present (West 1983). True steppes, on the other hand, are regarded as areas of grassland and herbaceous plants in which, apart from the areas of forest steppes, woody plants are absent.

10.5.2
Life-form and adaptation to winter cold and summer drought

Most of the herbaceous plants in the ecozone belong to the hemicryptophytes. A wide variety of spring geophytes and annuals whose buds or seeds survive the winter in or at the soil surface, are also present especially where a shallow snow cover gives protection from stress caused by cold.

In summer the stress from drought is also considerable. The number of buds forming in spring and early summer varies greatly, for example, from year to year, depending on the amount of precipitation. In the southern Russian short grass steppe, the above ground biomass ranges from 4.5 to 6.3 t ha^{-1} in rainy years and from 0.7 to 2.7 t ha^{-1} in dry years. The below ground biomass remains the same (Walter and Breckle 1986).

In the short grass and desert steppe, the drought stress is greater if the availability of water is limited further because of high salt content in the soil. Additional stress factors include over grazing by wild or domestic animals, large ranges in temperature and fire.

Owing to the limited precipitation, many plants are xerophytes. As aridity increases in the ecozone, leaves decline in size and are thicker, the epidermis and stomata cells become smaller, the number of stomata per leaf surface and the leaf vein density increase. The leaves of many species roll up as it becomes dryer and some lose their leaves, the latter can also be a reaction to cold.

Since the mineral nutrients are extracted from the soil with the water, this means that a reduced uptake of water directly correlates to a reduced absorption of nutrients, which consequently means shortcomings in two further respects: in the photosynthetic exchange of gas (due to the closure of leaf pores as a result of drought) and also in the secondary material synthesis (caused by a lack of minerals).

10.5.3
Animals

Large numbers of animals inhabit the steppes. Herds of various species of ungulates are, or were, present in all types of steppe. Wild horses and saiga antelopes in Eurasia, guanacos and pampas deer in east Patagonia and, in North America, bison, pronghorns and deer. In the 18th century, there were also large herds of mustangs, the descendants of the horses introduced by the Spanish in the 17th century. In areas of still intact steppe, there are high densities of hares, rabbits and rodents, including ground squirrels, hamster, prairie dogs, guinea pigs and numerous varieties of mouse. The ungulates and some rodents are herbivores, many rodents are also omnivores. Both are responsible for much of the turnover in the steppe ecosystem. As long as they do not over graze, their feeding stimulates the primary production to a higher biomass formation and also accelerates the subsequent recycling. Characteristic for all small rodents are a periodic mass propagation at intervals of several years which, during its peak, can result in up to 90% of the plant mass being consumed in an area.

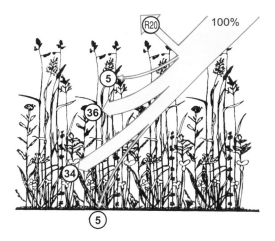

Fig. 10.4. Distribution of solar radiation in a meadow. Light penetrates into grassland further than into forests. More than 50% of the light reaches the middle of the grass layer. Source: Cernusea 1975

Some species of birds, grouse and bustards for example, were, and in some areas still are, important because they feed on grass seeds.

Grasshoppers, which can consume up to 25% of the shoot production are the most important herbivores among the invertebrates. Several varieties of beetle, including weevils, and butterfly caterpillars are also significant consumers.

Among the carnivores, coyote, badgers, weasels and a variety of birds of prey (eagles, buzzards, kites and falcons) are present in large numbers, a reflection of the high density of small herbivores available for prey.

10.5.4
Biomass, primary production and decomposition

Despite the small *biomass, primary production* in the steppes is very high with ranges from 2 to 15 $t\,ha^{-1}\,a^{-1}$, depending on the steppe type, Where there are woodland and tall grass steppes in which the woods and grass areas differ only in their soil type, the primary production is similar in both although the biomass in the woodlands can be 10 to 15 times greater. The steppes are more productive because they have only photosynthetically active organs at the surface and do not develop unproductive woody stems above ground, although, compared to trees, the ratio of roots to shoots is higher. The distribution of light in the grass layer is also favorable (Fig. 10.4) since the mostly vertical or near vertical blades of grass allow at least half the radiation usable for photosynthesis to reach the central area of the grass cover. In woodland, often only about 10% of the light reaches the trunk area.

In all areas of grassland, including those in the tropics, in which the water supply is temporarily at a minimum, the regional volume of above ground biomass and primary production can be correlated with the mean annual

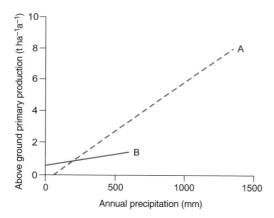

Fig. 10.5. The relationship between above ground primary production and annual precipitation in the North American steppes. A shows the variations in production at different locations in the central grasslands of the U.S.A. in relation to mean annual precipitation. B shows changes in production at one location in northern Colorado over a long period in relation to variations in mean annual precipitation. B increases more slowly than A because the reaction of the vegetation structure to precipitation surpluses in one year is limited and delayed in effect over several years. There is no relationship between NPP and temperature differences in individual years. Source: Lauenroth and Sala 1992

or the growing season precipitation, or the length of the growing season or actual evaporation (Figs. 10.5, 13.12 and 14.10). This correlation is valid for the same location with a varying precipitation over a number of years and also for different locations with varying long term annual means. In the North American steppes, about 5 kg of above ground biomass per hectare is produced per millimeter of annual precipitation (Risser 1988).

Since the shoot biomass dies off at the latest in the autumn, the amount of litter supplied each year is about as high as the above ground NPP of the same year. This litter is then rapidly decomposed, mainly within one year, by a rich soil flora and macro and micro soil fauna. Therefore, deep litter layers do not develop.

The lifetime of the below-ground phytomass is sometimes longer; being the maximum about 4 years for individual roots. This means that the root mass is also turned over relatively quickly. Thus, for the steppe ecosystem the unique case arises where (1) an extraordinarily short, almost 1-year metabolic cycle and energy flow exist and, correspondingly, (2) very approximately steady-state conditions prevail (Fig. 10.6).

In all other ecosystems, including the tundra and desert, the energy and minerals are hoarded in the form of long lasting woody growth and, or, waste that decomposes slowly. Only when plant stands have reached their aging phases, following phases of extreme conditions, such as those created by fire, wind storms or a drought, is the rate of decomposition greatly accelerated.

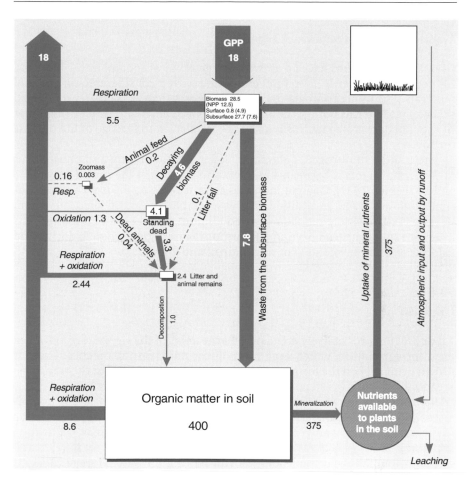

Fig. 10.6. Model of ecosystem in steppes. Based on data from a location near Matador in Canada in the Canadian praries for the period 1968–1972. Broken line = estimates. Mean shoot mass = 0.8 t ha^{-1}. Above ground NPP = 4.9 t ha^{-1} a^{-1}. 1.3 t ha^{-1} of the annual shoot production is lost in the decomposition process of the standing dead and does not reach the litter layer and heterotrophic organisms in the soil. Characteristic for the steppes is 1. much greater root mass compared to shoot mass, 2. material energy flows are high, both absolutely and in relation to the biomass and 3. humus forms the largest proportion of the organic matter in the ecosystem. Widths of arrows, areas of circle and of boxes are approximately equal in proportion to the volumes involved. Organic substances in t ha^{-1} or t ha^{-1} a^{-1}, minerals in kg ha^{-1} a^{-1}. Source: Coupland and Van Dyne 1979

10.5.5
Available supply and turnover of minerals

Grassland ecosystems have an above average mineral content in the biomass. All organic turnover is, therefore, combined with a considerable turnover of minerals. The mineral circulation in the tall grass steppes is greater than in all other zones in terms both of the amounts involved and the rate of throughflow. In the short grass steppes this is also true in relation to the size of the organic turnover.

The content of nitrogen, potassium, calcium, magnesium and phosphorus in the living shoots is between 4 and 5% and 2 to 3% in the roots (Titlyanova and Bazilevich 1979). Content increases with increasing aridity and salt concentration in the soil. In addition, in grasses in particular, silicon content and, in salty locations, sulphur, chlorine and sodium content are also higher. The litter and standing dead have smaller proportions of potassium, chlorine and sulphur but a higher content of silicon, iron and aluminum.

10.6
Land use

All dry regions are sparsely populated. Large areas of the steppes are, however, used for agriculture, either large farms growing grains or ranches. Grain is grown primarily on the former areas of tall grass steppes and in the transition area to the short grass steppes.

Ranching predominates in the short grass and desert steppes. The boundaries limiting grain cultivation lie at about 300 to 350 mm annual precipitation in the warmer steppes and 250 to 300 mm in the cooler steppe areas. Cultivation is possible at lower levels of mean annual precipitation if the seasonal distribution of rainfall is favorable and the soil has a high water storage capacity, also if drought resistant varieties of grain are grown and special dry farming techniques are used.

10.6.1
Large scale grain cultivation

Wheat is the most important market crop in areas of large scale *grain farming*. Farm enterprises are large, capital intensive, highly mechanized and require a minimum of labor. This highly commercial production reduces the costs so much that wheat cultivation could be extended into the former extensive pasture areas of the dryer steppes. Wheat produced in the steppes, including the steppes of the subtropics, makes a very large contribution to the supply of food in the world. The natural advantages of high soil fertility, high solar radiation, and the relatively level surfaces all favor the use of large agricultural machinery and consequently the management of production in large units.

Dry farming techniques and *irrigation* are used in the boundary areas where precipitation is insufficient for crop cultivation unless drought resistant crops

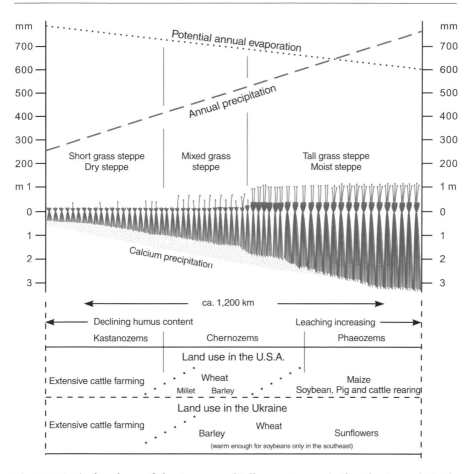

Fig. 10.7. Agricultural use of short grass and tall grass steppes in the Ukraine and North America. Large scale grain cultivation predominates in the former mixed grass praries and in the dry boundary areas of the former tall grass steppes. Ranching dominates in the short grass steppes. The American corn belt is in the area of tall grass steppe. Source: Jätzold 1984

such as millet, peanuts, chick peas or sesame are grown. In areas of dry farming, fields are left fallow for a year so that evaporation is reduced and water stored in the soil for future years. Depending on the precipitation deficit, fields are left fallow every second, third or fourth year so that half, one third or one quarter of the cutivated area is fallow in any one year. Figure 10.8 shows the variation in the annual harvest in relation to the period of fallow. Cattle pastures growing clover are another form of fallow which also leads to an increase in water supply in the soil after a few years.

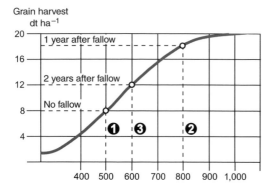

Grain harvest
dt ha⁻¹

① Annual cultivation without fallow
Annual available water = 500 mm
Yield 8 dt ha^{-1}a^{-1}

② Annual alteration between cultivation and fallow
Available water in cultivation year = 800 mm
Yield $(18 + 0)/2 = 9$ dt ha^{-1}a^{-1}

③ Two years cultivation, one year fallow
Available water in 1st cultivation year = 800 mm
in 2nd cultivation year = 600 mm
Yield $(18 + 12 + 0)/3 = 10$ dt ha^{-1}a^{-1}

Fig. 10.8. The effect of fallow periods on dry farming systems. The highest yields are obtained in cultivation cycle 3. Annual available water in mm = annual precipitation plus soil water stored from previous years (annual precipitation is assumed to be 500 mm). Source Andreae 1983

10.6.2
Extensive pasture economy

In the arid lands of the earth, this is practised either in the form of (semi-) *nomadic herding* or as *ranching*. The former is the traditional form of pastoralism in the arid lands of the Old World, from the deserts to the steppes or savannas. Nowadays, it is found mainly in the tropics/subtropics, and it will therefore be dealt with in the chapter on that ecological zone (see Chap. 13.6.1).

Ranching by contrast is the modern commercial form of an extensive pasture economy. It was developed first in the Americas and Australia by European settlers and spread later to other parts of the world, including southern Africa. Ranching competes with arable farming and has been increasingly supplanted by it and forced to move to dryer regions. Arable farming is competitive when the annual precipitation allows an annual pasture yield that supports a mean stocking density of 30 to 40 cattle per hectare of pasture.

Some of the following factors are typical for ranching.

1. Very large units of production from 500 to 100,000 hectares. Unit size increases with increasing aridity and therefore lower productivity of the pasture.

2. Cattle predominate with sheep in dryer areas. In some areas game are kept on ranches.

Table 10.1. Cattle densities in the western United States related to mean annual precipitation

Annual precipitation (mm)	Number of cattle per 100 hectares
< 250	3 to 5
250–500	5 to16
500–750	16 to 50

3. The primary product is beef cattle, usually of a single species or breed.

4. Grazing is controlled in large fenced enclosures. The fences are often the only indication that an otherwise natural region is used for ranching.

5. Ranching can be a high risk enterprise because of feed shortages caused by drought.

6. The carrying capacity, labor requirements, capital input and yields are low per spatial unit, the productivity of a spatial unit among the nomads is even lower. Productivity per unit of labor is, however, very high.

7. The investment required to set up a ranching unit is high.

The optimal carrying capacity of pasture possible without damage to the resource can be estimated from the above ground primary production of suitable fodder plants. Maximum output is limited by precipitation (Fig. 10.5). In most cases the rain use efficiency is between 3 and $6\,kg\,ha^{-1}\,a^{-1}$ dry matter per millimeter of mean annual precipitation. Table 10.1 shows the approximate relationship between the pasture requirement per cattle unit and the distribution of mean annual precipitation in the western United States (Andreae 1988).

With an increase in the number of cattle per unit of fodder supply, there may be a change in the steppe flora if the cattle show a preference for particular species. This could lead to a selective defoliation, or a varying sensitivity by the plant species in response to defoliation could put certain plant species at a disadvantage. The feed value of the plant production and consequently the capacity of the pasture are then considerably reduced.

In addition, the tread of heavy cattle on intensely grazed pastures leads to compression of the soil and a reduction in the possible infiltration rate of water. Overland flow and gully erosion may follow. Rain splash can also seal the soil surface.

Problems associated with ranching do not occur to the same extent if herds of game are kept commercially. Other difficulties have, however, limited development in the steppes to only a few projects, bison in North America and lamas in Patagonia are examples.

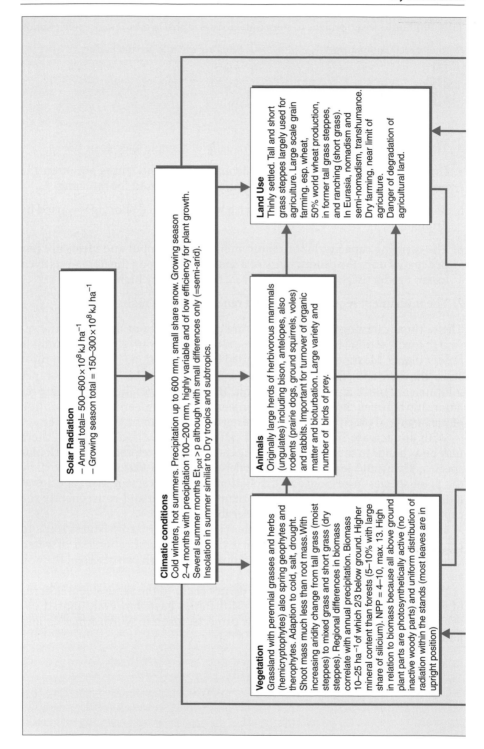

Solar Radiation
- Annual total= $500-600 \times 10^8$ kJ ha^{-1}
- Growing season total = $150-300 \times 10^8$ kJ ha^{-1}

Climatic conditions
Cold winters, hot summers. Precipitation up to 600 mm, small share snow. Growing season 2–4 months with precipitation 100–200 mm, highly variable and of low efficiency for plant growth. Several summer months Et$_{pot}$ > p although with small differences only (=semi-arid). Insolation in summer similiar to Dry tropics and subtropics.

Vegetation
Grassland with perennial grasses and herbs (hemicryptophytes) also spring geophytes and therophytes. Adaption to cold, salt, drought. Shoot mass much less than root mass.With increasing aridity change from tall grass (moist steppes) to mixed grass and short grass (dry steppes). Regional differences in biomass correlate with annual precipitation. Biomass 10–25 ha^{-1} of which 2/3 below ground. Higher mineral content than forests (5–10% with large share of silicium). NPP = 4–10, max. 13. High in relation to biomass because all above ground plant parts are photosynthetically active (no inactive woody parts) and uniform distribution of radiation within the stands (most leaves are in upright position)

Animals
Originally large herds of herbivorous mammals (ungulates) including bison, antelopes, also rodents (prairie dogs, ground squirrels, voles) and rabbits. Important for turnover of organic matter and bioturbation. Large variety and number of birds of prey.

Land Use
Thinly settled. Tall and short grass steppes largely used for agriculture. Large scale grain farming, esp. wheat, 50% world wheat production, in former tall grass steppes, and ranching (short grass). In Eurasia, nomadism and semi-nomadism, transhumance. Dry farming, near limit of agriculture. Danger of degradation of agricultural land.

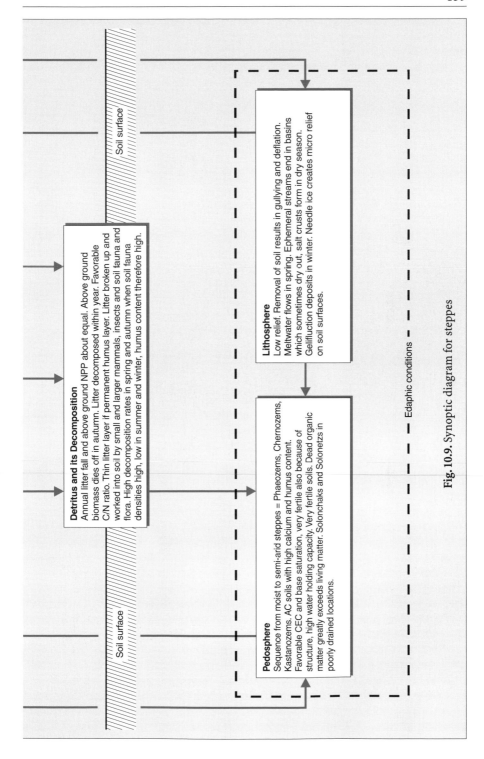

Detritus and its Decomposition

Annual litter fall and above ground NPP about equal. Above ground biomass dies off in autumn. Litter decomposed within year. Favorable C/N ratio. Thin litter layer if permanent humus layer. Litter broken up and worked into soil by small and larger mammals, insects and soil fauna and flora. High decomposition rates in spring and autumn when soil fauna densities high, low in summer and winter, humus content therefore high.

Lithosphere

Low relief. Removal of soil results in gullying and deflation. Meltwater flows in spring. Ephemeral streams end in basins which sometimes dry out, salt crusts form in dry season. Gelifluction deposits in winter. Needle ice creates micro relief on soil surfaces.

Pedosphere

Sequence from moist to semi-arid steppes = Phaeozems, Chernozems, Kastanozems. AC soils with high calcium and humus content. Favorable CEC and base saturation, very fertile also because of structure, high water holding capacity. Very fertile soils. Dead organic matter greatly exceeds living matter. Solonchaks and Solonetzs in poorly drained locations.

Soil surface

Soil surface

– – – Edaphic conditions – – –

Fig. 10.9. Synoptic diagram for steppes

CHAPTER 11

Subtropics with winter rain

11.1
Distribution

The Subtropics with winter rain covers over 2.5 million km^2 or 1.7% of the landmass and is the smallest and most fragmented of the ecozones. The five major areas of the zone are located on five continents (Fig. 11.1). All lie on the western side of a continent between 30° and 40° latitude north or south of the equator and between the Dry tropics and subtropics and Temperate midlatitudes. In general, they reach about 100 km inland from the coast. Only in the Mediterranean does the ecozone extend further eastwards into the landmass but is always near a coast. The ecozone also reaches its most northern extent, about 45°N, in the area of the Mediterranean. In part because of the fragmentation, the flora and fauna, ecological characteristics and cultural and economic development within the zone are greatly varied.

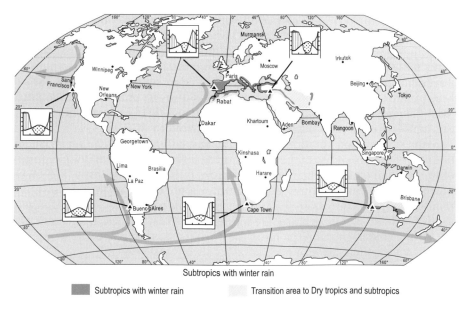

Subtropics with winter rain

■ Subtropics with winter rain   Transition area to Dry tropics and subtropics

Fig. 11.1. Subtropics with winter rain

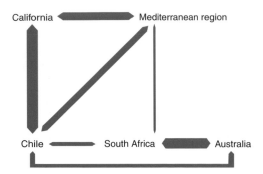

Fig. 11.2. Degree of affinity between the five areas of the Subtropics with winter rain. The arrow width is proportional to the degree of affinity; climate, vegetation, land use etc. have been included in the comparison. Source: Di Castri et al. 1981

11.2
Climate

In summer, the Subtropics with winter rain are influenced by subtropical-tropical highs and clear, sunny, dry weather is typical for the area. During the winter months periods of high pressure alternate with cloudy weather as belts of cyclones move from the midlatitudes towards the equator and bring precipitation associated with the fronts that pass through. Cold air bursts followed by frost also occur in winter, even in the lowlands, but these colder periods do not last long.

Mean annual precipitation increases in the direction of the poles to a maximum of 800 to 900 mm. The periods of rainfall are also longer than in the dryer areas. The summer can include some months with no precipitation at all. The boundary to the Temperate midlatitudes lies where the limitation to plant growth in summer due to lack of moisture is no longer noticeable. Towards the equator, the dry season lasts for a least seven months and annual precipitation is less than 300–350 mm. This is also the limit of the *sclerophyllous* phanerophytic vegetation which is characteristic of the zone, beyond are the subtropical shrub and grass steppes with winter rain.

Mean summer temperatures are lower at this latitude than in other ecozones because of the proximity of the coast and the often relatively low temperatures of the sea due to the cold currents and more frequent cloud cover (Fig. 11.1). Mean monthly temperatures exceed 18 °C in at least four months in most regions but are rarely more than 20 °C, except in some areas of the Mediterranean. In winter, temperatures are moderate, usually not less than a mean of 5 °C in the coldest month. In some of the border areas towards the poles means may be lower; there are also occasional frosts in most winters.

Low temperatures do not cause a prolonged pause in growth. The combination of temperature and rainfall in spring and autumn are more favorable for the vegetation and plant development than the winter months but the greatest stress occurs in summer when the limited availability of water is an important stress factor in plant growth.

11.3
Relief and drainage

The summer aridity of the ecozone limits fluvial and denudative geomorphological processes to the winter months when the effects can be considerable because of the combination of high relief energy and thin soil cover in large parts of the zone. Gaps in the vegetation which are present year-round or develop after the summer drought, together with a shallow litter layer, limit percolation and the ability of the surface to absorb water so that overland flow events are frequent. Fire, by reducing or destroying the vegetation cover and sometimes also the litter and humus at the surface increases the vulnerability to denudation. Fires occur regularly in areas of sclerophyllous shrub.

Fluvial erosion and wash denudation are more intensive and landslides more frequent where the plant cover has been damaged or almost destroyed by over grazing. This type of degradation of the vegetation is so widespread that it has become a characteristic of the entire ecozone.

Because such a high proportion of the precipitation reaches the streams as overland flow, stream flow is closely related to precipitation events and subject, therefore, to considerable variation. Small streams rapidly develop into torrents carrying large quantities of pebbles and suspended load, up to $> 50 \, \mathrm{kg \, m^{-3}}$ (Le Houerou 1981). Dams may be in danger of breaking and there is extensive flooding, deep erosion and uncontrolled deposition of debris cones or fans where the gradient of the streams is sharply reduced as they reach the plains. The fine material is deposited beyond the coarse material, usually near or on the coast and contributes to the extension of the fertile flood plains and deltas. Around the Mediterranean, alluvial areas have been of considerable importance for settlement and agricultural production and contrast with other more sparsely settled coastal regions. During the summer, stream flow is greatly reduced or ephemeral.

11.4
Soils

Soils vary greatly, often over short distances. Variations in precipitation, bedrock, soil erosion, karst development and flood damage and also soils that have resulted from paleoclimatic changes all play a role in the small scale pattern of soil distribution. Most soils are of low fertility because of lack of phosphorus and nitrogen, other, nutrient poor, soils have developed on Precambrian and Paleozoic rocks or quartz sands over large areas of South Africa and Australia (Fig. 11.3).

Chromic Luvisols are the most frequent soil type. They occur on moderate slopes and have developed undisturbed over a long period. Bright red to reddish brown in color, leached, and generally on limestones or dolomites, they are also rich in bases, poor in humus, highly erodable and shallow. During dry periods hardening takes place. Their reddish coloring is caused by a fine distribution of hematite which, together with a high clay content, decalcification of the

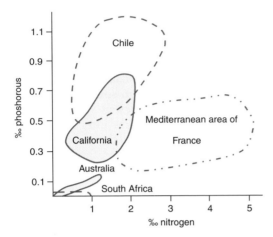

Fig. 11.3. Phosphorous and nitrogen content in the soil in the five areas of the Subtropics with winter rain. Soils in Australia and South Africa are particularly low in these nutrients. Source: Di Castri et al. 1981

upper soil horizon and a secondary layer of calcite enrichment in the subsoil, are indications of their long development. Kaolinite is also often present (Jahn 1997).

Much less widespread are *Chromic Cambisols,* also reddish in color. Around the Mediterranean, Chromic Luvisols were formerly known as terra rosa and Chromic Cambisols as terra fusca. They are thought to have been developing since the Tertiary, or certainly since the Pleistocene.

Chromic Luvisols are common in California, central Chile and the Cape area of South Africa, fairly common around the Mediterranean but hardly present in the Australian areas of the ecozone. Calcisols, with a secondary calcium enrichment, developed particularly in Australia and the Mediterranean, and Eutric Cambisols are the other most frequently occurring soils in the zone.

11.5
Vegetation and animals

11.5.1
Sclerophyllous vegetation

The Subtropics with winter rain have the second highest number of species of all ecozones, although the zone covers the smallest area. Many are endemic species. The small area in South Africa with winter rain has the largest number of species per spatial unit. There are over 6,000 vascular plants in this region, three times as many as in similiar areas of tropical rainforest. In the southwest Australian and Californian areas of the zone, approximately 5,000 plants have been identified, about one quarter of all North American species (Mooney 1988). Others have estimated a total of 8,000 plants for these areas (Hobbs

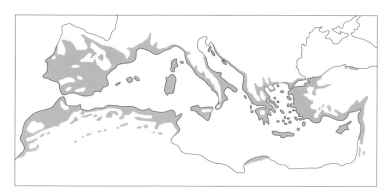

Fig. 11.4. Distribution of maquis and garrigue in the Mediterranean region. Source: Quezel 1981

1992), depending on how borders are defined. From 18,000 to 25,000 species grow in the Mediterranean region alone, half of which are endemic.

With the exception of the dryest areas and those poorest in nutrients, the entire ecozone was probably covered with *evergreen sclerophyllous forest* and, in the Northern Hemisphere, also pine forest. Three types of evergreen oak forest were also significant in the Mediterranean area, the live oak (Quercus ilex) and the cork oak (Quercus suber) in the west and in the eastern Mediterranean area, the Quercus calliprinos forest.

Thousands of years of intervention by man around the Mediterranean has all but destroyed the natural forests. The *sclerophyllous shrubs* that largely replaced the forest have become synonymous for the landscape of much of the area and a criterion for its definition (Fig. 11.4).

Maquis is the term used to cover the wide variety of vegetation types with taller growth that have developed in this climate. Other regional names are chaparral in California, matorral in spanish speaking areas, macchia in Italy, mallee in Australia and fynbos in South Africa.

Garrigue is the term used to describe the lower growth vegetation. Regionally, tomillares (Spain), phrygana (Greece), renosterveld (Africa), kwongan (Australia) coastal sage (North America) and jaral (Chile) are used to describe this type of low shrubby cover.

The taller maquis ranges in height from about half a meter to at most 2 to 3 meters. A dense stand of a variety of shrubs interspersed with small trees is typical for the vegetation form with green barks.

Many of the shrubs have green barks and are either leafless or have small leaves. Some have thorns. Dwarf and low shrubs grow in the undergrowth and there is a dense herbaceous cover in clearings.

Sometimes garrigue consists of quite dense stands of chamaephytes between which are herbaceous plants and geophytes. If the land is is no longer pastured or burnt over, higher shrubs invade together with hemicryptophytes, perennial grasses and herbaceous plants, in their shade. The natural regeneration of the vegetation, after an initial cover of grasses and herbaceous plants, leads to

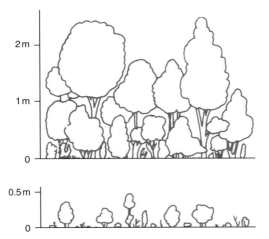

Fig. 11.5. Structure of a high growth dense maquis and a low growth open garrigue. Source: Tomaselli 1981

a cover of sclerophyllous shrubs, providing cold and aridity are not too severe. Depending on the degree of interference by man, eventually sclerophyllous forests or pine forests develop, although under semi-arid warm conditions, sclerophyllous shrubs tend to become the climax community. Vegetation in these regions can be designated either as climax vegetation of a particular degree of aridity, as semi-permanent replacement communities following interventation, as relatively short-lived post forest indicators or even as post agricultural indicators.

11.5.2
Adaptation to summer drought

Unlike the subtropics and tropics with summer rain, the Subtropics with winter rain are dominated by evergreen trees and shrubs whose leaves have a high proportion of structural tissue containing cellulose and lignin, and are also relatively thick, stiff and leathery (Fig. 11.6). Even with major water loss, the leaves do not whither.

The combination of characteristics present in sclerophyllous plants is represented in a wide variety of plant families. They are an example of environmental convergence that has followed drought stress and high solar radiation, conditions that are particularly frequent in the Subtropics with winter rain. However, it is also found elsewhere, under conditions of nitrogen shortage.

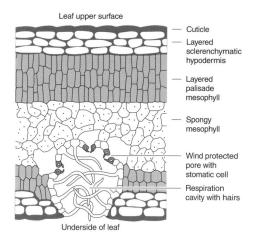

Fig. 11.6. Scleromorphic leaf of Nerium Oleander with a thickened hypodermis, mesophyll and sunken stomata. Source: Larcher 1994

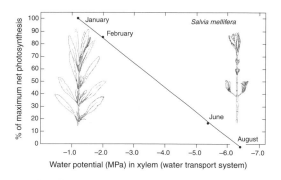

Fig. 11.7. Seasonal variations in the net assimilation of Salvia mellifera with dimorphic leaves in chaparral in California U.S.A. Source: Mooney and Miller 1985

The nature of the leaves in sclerophyllous plants helps to control their water budgets. Typically they have a thickened epidermis, a shiny waxy surface, hairs, narrow veins and a low pore area, the pores are dense but small.

Some plants are seasonally dimorphic in that the leaves developed during the rainy season are replaced in the dry season by smaller, fewer, more xeromorphic leaves which reduce the transpiration surface and amount of water given off by the leaves and plant. The larger leaves of the rainy period have double the photosynthesis rate of sclerophyllous plants (Margaris and Mooney 1981), the smaller leaves a considerably lower rate, but nevertheless even with a low water potential, still have a net gain (Fig. 11.7). Many species of shrub in the garrigue are diomorphic in this way.

Although sclerophyllous trees and shrubs cover large areas and are characteristic of the zone, they are not the most frequent growth forms. Hemicrypto-

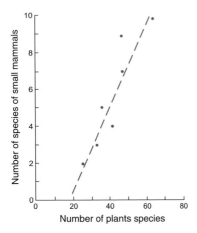

Fig. 11.8. Correlation between the number of species of small mammals and plant diversity in ecosystems of the southern Australian mediterranean area. Source: Specht 1994

phytes, therophytes and geophytes are more numerous both in terms of species and individual plant numbers. Many winter annual and perennial herbaceous plants flower in spring. Succulents are also widespread in California and Chile. The agaves and prickly pear often seen in the Mediterranean area are introduced.

11.5.3
Animals

The variety of plants, orographic differentiation of precipitation and finely meshed pattern of shrub, heaths, grass and woodland provide a wide range of habitats for fauna. There are large numbers of bird species including song birds, birds of prey, gallinaceous birds and doves, reptiles, especially lizards, also ants, spiders, centipedes, millipedes, beetles, butterflies, termites, scorpians and other invertebrates. The number of species increases with the increase in the number of vascular plants and also of rainfall. An exception are the lizards whose numbers decline as the density of the canopy increases (Specht 1994).

The ecozone also attracts animals from the neighboring semi-deserts. From the midlatitudes such migrations are rare, with the exception of migratory birds that use the region to rest while passing through.

11.5.4
Fire

Shrub and forest fires are a frequent occurrence, generally reported annually from California, the Mediterranean, South Africa and Australia. The mean period of recurrence in most Mediterranean regions is only a few tens of years.

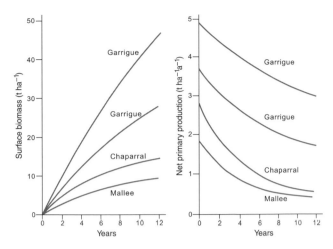

Fig. 11.9. Changes in the biomass and primary production in selected sclerophyllous shrubs in the twelve years following a fire. The garrigues are near Montpellier, France, the chaparrals near San Dimas, California, U.S.A. and the mallee near Keith, South Australia. As the succession initiated by the fire continues, the NPP is reduced each year and the increase in biomass slowed down. Source: Specht 1981

Fires are the oldest recurring events of the ecosystem, although today they are often caused by man. Sclerophyllous vegetation is particularly prone to fires. Heat and dryness at the same time of year, dense stands of trees and shrubs, the etheric oils and resins in the leaves of this type of vegetation combine to create a highly flammable cover. Shrub and woodland fires can destroy the entire surface plant mass and are far more devastating than the grass fires of the winter dry tropical savanna.

Since woodland and shrub fires are a natural environmental factor in the Mediterranean, many plants have adapted to periodic burning, by either regenerating very rapidly, or by improving their ability to germinate their seeds after a fire, or even germinating only when a fire has occurred.

Many shrubs are not only adapted to fire but require fires to ensure regeneration. Fire is an ecological factor and older communities can be termed fire climax communities. They are part of a succession which does not become a climax community within the broad climatic conditions of a region but is always returned to an earlier stage by fire from which it progresses again in the succession.

An advantage of burning over an area is that the mineral nutrients in the organic matter are released earlier than if natural biological and chemical decomposition of the organic waste had taken place. There is also then an increase in biomass which peaks in the first years after a fire. The reduction in the annual increment after this peak is paralleled by a decline in surface productivity (Fig. 11.9).

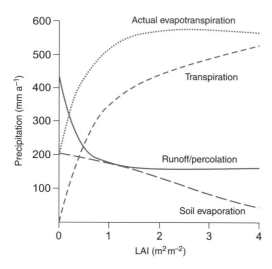

Fig. 11.10. Changes in the water balance in a Quercus coccifera garrigue area near Montpellier, France after fire and regeneration. The soil is bare of vegetation after the fire. The leaf area index (LAI) = 0. Two thirds of the precipitation (425 mm a^{-1}) percolates into the ground, one-third evaporates. At the end of the first regeneration phase the LAI < 1 and the loss from runoff and percolation is reduced by half. When the LAI = 1–4, losses reach < 150 mm a^{-1} and remain at this level. Evapotranspiration increases to > 500 mm a^{-1} and the transpiration proportion increases to 90% with an LAI of 4. Evaporation from the soil is reduced from > 200 mm to 50 mm a^{-1}. Source: Rambal 1994

Overland flow and infiltration may also increase on burnt over slopes, causing greater soil erosion and the removal of nutrients (Fig. 11.10). The sediment removed during erosion is deposited at the foot of the slope. The more frequent the fires the greater the degradation.

11.5.5
Biomass and primary production

The production capacity of the vegetation in the Subtropics with winter rain is limited by the lack of water during the warm season and by a moderate lack of warmth during the rainy season. Production in the ecosystem is, therefore, relatively low in relation to the length of the growing season with the highest growth rates occurring in the spring. Many woody plants are sclerophyllous and evergreen which allows growth to continue during the dry season at a lower level; some also have the ability to produce immediately should occasional rainfall occur during a dry period. Year-round photosynthesis is possible although production rates even with optimal moisture supply do not equal those of deciduous trees or shrubs in areas of winter precipitation. Deciduous vegetation has a limited production in the dry season, but often has a higher annual production than the sclerophyllous and evergreen plants because of growth during the winter rains.

Table 11.1. Production characteristics of selected mediterranean plant formations

Plant formation	Evergreen oak forest	Evergreen shrubs		Subshrub Phrygana
		Chaparral	Garrigue	
Research area	France (Le Rouquet)	California	France (St. Gély)	Greece
Age of stand (yrs)	150	17–18	17	–
Height (m)	11	≈ 1.5	0.8	< 1
Leaf area index (LAI) $(m^2\,m^{-2})$	4.5	2.5	–	1.7
Biomass $(t\,ha^{-1})$				
Shoot mass	269	20.39	23.5	10.95
Branches	262	16.72	19.5	8.86
Leaves	7	3.67	4.0	2.09
Root mass	≈ 50	≈ 12.23	–	16.18
Total	319	32.62	–	27.13
Root/shoot relationship	0.19	0.60	–	1.48
Primary production $(t\,ha^{-1}\,a^{-1})$				
Growth above ground	2.6	1.30	1.1	2.02
Litter fall	3.9	2.82	2.3	2.10
Total above ground production	6.5	4.12	3.4	4.12

Source: from Mooney 1981

The structural characteristics and the length of the growing season influence the productivity of the vegetation (Table 11.1). The values in the evergreen oak forest are highest where the biomass and leaf area index, in this case 319 t ha^{-1} and 4.5 respectively, are high and the root-shoot ratio, 0.19, is correspondingly low. By contrast the Greek phrygana (garrigue) has a biomass of 27 t ha^{-1} and LAI of 1.7, together with a higher root-shoot ratio of 1.48. Dryness and an environment more affected by man underlie these values.

The annual net primary production in the oak forest is 6.5 t ha^{-1} and in the phrygana 4.12 t ha^{-1}. In the garrigue and chaparral in California, the NPP is 3.4 t ha^{-1} and 4.12 t ha^{-1} respectively. The plant formations in the Subtropics with winter rain have, therefore, a relatively wide range but are generally below those in the Temperate midlatitudes (Fig. 11.11). The energy fixed in the annual primary production ranges from only 0.17 to 0.3% of the annual solar radiation. The restricted water supply and deficits in nutrient supply in the soil are the two most important limiting factors for plant growth.

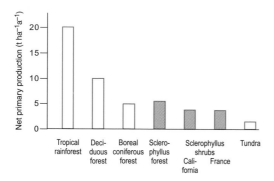

Fig. 11.11. Net above ground primary production in a sclerophyllous mediterranean formation compared to other formation types. The sclerophyllous forest produces similiar quantities to the boreal forest but less than the temperate deciduous forest which receives less solar radiation than the Mediterranean area. Source: Mooney 1981

11.6
Land Use

The coastal location and the long hours of sunshine in summer have been a factor in the devlopment of many areas of the ecozone where seafaring, fishing and in recent decades, tourism are major economic activities. Of greater significance, especially during the last 50 years, has been the advantage of a moist winter climate which allows crops, particularly vegetables and fruit to be harvested in winter and spring and marketed in cooler climates. Most of the areas of the ecozone have developed an extensive world trade in agricultural products in recent years. Precipitation in winter is everywhere sufficient for some form of cultivation but has to be supplemented by irrigation in the summer months.

Plants that originated in the Temperate midlatitudes dominate the commercial market. Winter wheat, barley, potatoes, and field vegetables such as salad, onions, tomatoes, cauliflower, artichokes, aubergine and broccoli as well as maize are all widely grown. Around the Mediterranean, winter wheat, sown in September is harvested in May. Irrigation is necessary in summer for both field vegetables and crops such as rice and cotton which require heat but are also cold sensitive.

Olives, grapes, apricot, peach, citrus and other fruit trees are cultivated in almost every region of the zone. Arable farming is primarily on the level ground near the coast and in the valleys of the Mediterranean area. Vineyards, olive groves and orchards are planted on the slopes. Higher up, natural pastures provide grass for sheep and goats. Around the Mediterranean, shepherds move up in summer to the high pastures in a form of transhumance, sometimes covering long distances with their flocks (Fig. 11.12). There has been a decline in this practice in recent years but it is still continued in remote areas.

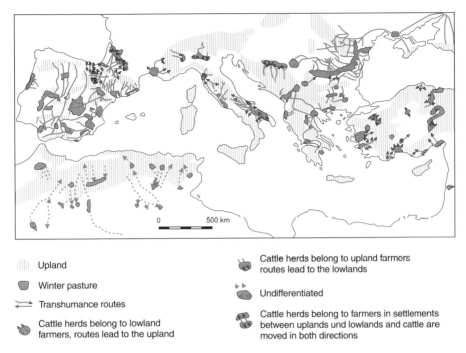

Upland

Winter pasture

Transhumance routes

Cattle herds belong to lowland
farmers, routes lead to the upland

Cattle herds belong to upland farmers
routes lead to the lowlands

Undifferentiated

Cattle herds belong to farmers in settlements
between uplands und lowlands and cattle are
moved in both directions

Fig. 11.12. Transhumance in the Mediterranean area. Source: Grigg 1974

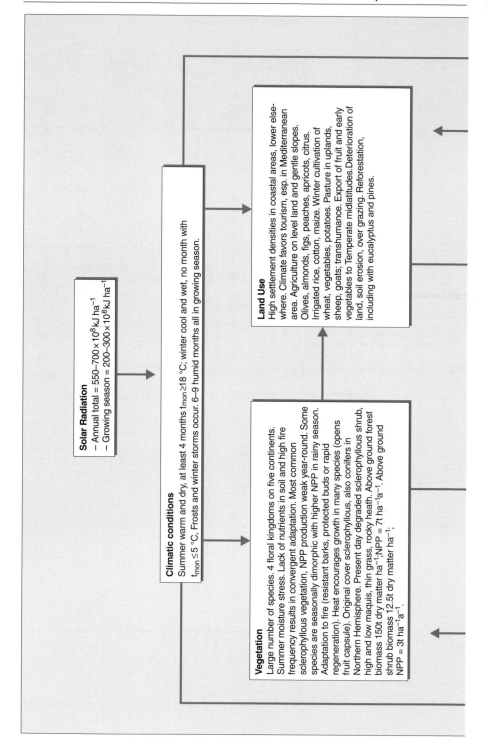

Solar Radiation
- Annual total = $550-700 \times 10^8$ kJ ha^{-1}
- Growing season = $200-300 \times 10^8$ kJ ha^{-1}

Climatic conditions
Summer warm and dry, at least 4 months $t_{mon} \geq 18$ °C; winter cool and wet, no month with $t_{mon} \leq 5$ °C, Frosts and winter storms occur. 6–9 humid months all in growing season.

Land Use
High settlement densities in coastal areas, lower elsewhere. Climate favors tourism, esp. in Mediterranean area. Agriculture on level land and gentle slopes. Olives, almonds, figs, peaches, apricots, citrus. Irrigated rice, cotton, maize. Winter cultivation of wheat, vegetables, potatoes. Pasture in uplands, sheep, goats; transhumance. Export of fruit and early vegetables to Temperate midlatitudes.Deterioration of land, soil erosion, over grazing. Reforestation, including with eucalyptus and pines.

Vegetation
Large number of species. 4 floral kingdoms on five continents. Summer moisture stress. Lack of nutrients in soil and high fire frequency results in convergent adaptation. Most common sclerophyllous vegetation, NPP production weak year-round. Some species are seasonally dimorphic with higher NPP in rainy season. Adaptation to fire (resistant barks, protected buds or rapid regeneration). Heat encourages growth in many species (opens fruit capsule). Original cover sclerophyllous, also conifers in Northern Hemisphere. Present day degraded sclerophyllous shrub, high and low maquis, thin grass, rocky heath. Above ground forest biomass 150t dry matter ha^{-1};NPP = 7t ha^{-1}a^{-1}. Above ground shrub biomass 12.5t dry matter ha^{-1}; NPP = 3t ha^{-1}a^{-1}.

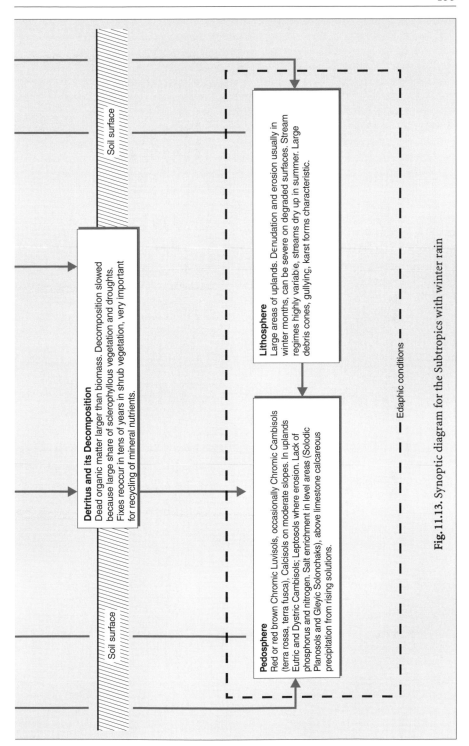

Detritus and its Decomposition
Dead organic matter larger than biomass. Decomposition slowed because large share of sclerophyllous vegetation and droughts. Fixes reoccur in tens of years in shrub vegetation, very important for recycling of mineral nutrients.

Soil surface

Lithosphere
Large areas of uplands. Denudation and erosion usually in winter months, can be severe on degraded surfaces. Stream regimes highly variabe, streams dry up in summer. Large debris cones, gullying, karst forms characteristic.

Pedosphere
Red or red brown Chromic Luvisols, occasionally Chromic Cambisols (terra rossa, terra fusca), Calcisols on moderate slopes. In uplands Eutric and Dystric Cambisols; Leptosols where erosion. Lack of phosphorus and nitrogen. Salt enrichment in level areas (Solodic Planosols and Gleyic Solonchaks), above limestone calcareous precipitation from rising solutions.

Edaphic conditions

Fig. 11.13. Synoptic diagram for the Subtropics with winter rain

CHAPTER 12

Subtropics with year-round rain

12.1
Distribution

The Subtropics with year-round rain are also present on five continents (Fig. 12.1) and lie between 25° and 35° latitude on the eastern side of North and South America, Asia, Africa and Australia. Their area totals about 6 million km^2, approximately 4% of the landmass Towards the equator they are bordered by the Tropics with year-round or with summer rain and polewards by the Temperate midlatitudes.

The boundaries can be defined by temperature (Hübl 1988). The limit of frost or the 18 °C isotherm for the coldest month in the lowlands form the borders of the ecozone towards the tropics. In the direction of the midlatitudes, 4 or at most 5 months with a monthly mean of at least 18 °C and a mean for the coldest month of at least 5 °C, exceptionally 2 °C in more continental locations, define the border areas. The growing conditions in the ecozone are also affected by variations in temperature, although less than in the midlatitudes.

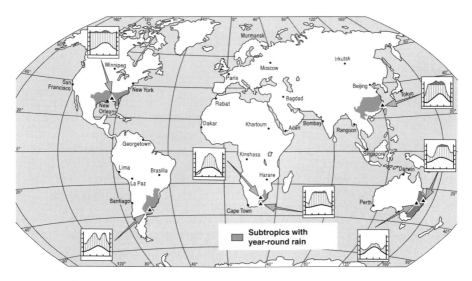

Fig. 12.1. Subtropics with year-round rain

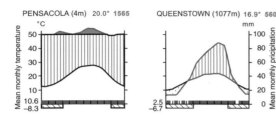

Fig. 12.2. Climates at two stations in the Subtropics with year-round rain. Pensacola, Florida, U.S.A. has a high year-round precipitation with summer maximum. Winter temperatures are lower but usually remain above 5 °C. Frost can occur (hatched bars along x-axis). Queenstown, South Africa lies 150 km inland and is an example of a transition area. Percipitation falls mostly in the summer, but the winter is still subhumid.

A transition area some 100 km wide exists west of the Subtropics with year-round rain where they merge with the dryer subtropical continental interiors.The annual precipitation in this transition area decreases westwards as, first in the winter months and then in the summer months, rainfall declines. The westerly limits of the ecozone are generally defined as the area in which there are fewer than five months with precipitation of (p [mm] $\geq 2t_{mon}$ [°C]) and thorn steppe replaces grass and woodland steppe. This threshold value is based on the fact that areas with at least five "humid" months also receive moderate amounts of precipitation during the "dry season" (in the climatic diagrams, the precipitation curves run just below the temperature curves during the "dry months"; Figs. 12.1 and 12.2), i.e. there is no true (i.e. rainless) dry season, but instead subhumid or semi-arid periods alternate with the humid periods. Many drought-adapted species of plant are able to grow all the year round under these conditions and for them the climate is within certain limits continuously humid.

12.2
Climate

Precipitation is year-round in the zone. During the summer months it is often substantial and nearly all regions have winter precipitation as well. Extended dry periods are infrequent and there is a general pattern of periods of low precipitation separated by periods of heavier rainfall. In the summer in both hemispheres low pressure heat cells develop over the continents which draw moisture laden air onshore from the east. Convection processes over the landmasses are followed by very heavy rainfall, often as showers. This is the basis for the summer maximum. Inland from the coasts the air masses become dryer and annual precipitation declines.

In the Northern Hemisphere in central North America and central Asia winter is associated with high pressure and cold air masses which bring both rain and snow. The snow is rarely deep or long lasting. Mean monthly temperatures are above 5 °C although temperatures can drop sharply under the influence

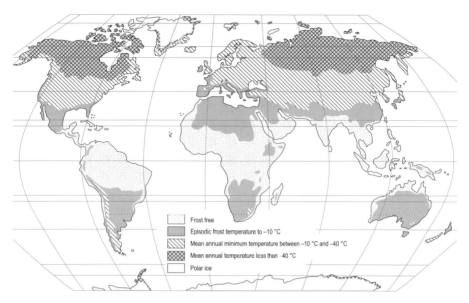

Fig. 12.3. Worldwide distribution of areas of frost. The Subtropics with year-round rain have episodic frost with temperatures falling to −10 °C. The damage to vegetation and tree crops (citrus) can be considerable because many subtropical plants are less well adapted to frost than plants in areas with periods of frost each year. Source: Larcher and Bauer 1981

of the continental Arctic air for short periods. Plant growth is more limited by cold spells than in the Subtropics with winter rain, but most plants do not become dormant for any length of time. Frost periods in nearly every winter prevent the cultivation of frost sensitive winter crops, also the growth of some moderately frost sensitive trees such as citrus is affected if temperatures drop suddenly.

Summers are hot because of the high insolation and summer temperatures are comparable to those in the tropics.

12.3
Relief and drainage

No geomorphological process is typical for the region. Processes common in the Tropics with year-round rain or the Temperate midlatitudes occur in different parts of the zone. There has been deep chemical weathering, although not as advanced as in some areas of the tropics. Gullying takes place to a greater extent than in the tropical rainforest because the forest cover is not as dense or as highly developed in the subtropics. Cyclonic storms, known as hurricanes in the Americas and typhoons in east and southeast Asia, reach the coastal areas (Fig. 12.4), (Boose et al. 1994). They bring high winds, up to at least 150km an hour, and intense precipitation events of over 100 mm ha^{-1}. Although they do

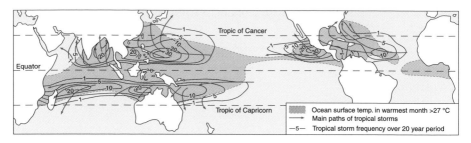

Fig. 12.4. Distribution and frequency of tropical storms (hurricanes and typhoons). Tropical storm paths lie across the Subtropical zone with year-round rain in the Northern Hemisphere. Source: Reading et al. 1995

not occur often, when they do such storms can have a major impact, sometimes reaching relatively far inland and causing soil erosion, crop damage and the destruction of buildings and vegetation.

12.4
Soils

Many of the soils in the ecozone are similar to the soils of the Tropics with summer and with year-round rain. *Acrisols* are the most widespread soil in the zone. They are leached and have a clay enriched argic B horizon and a low CEC of less than $24\,cmol(+)\,kg^{-1}$ clay and a base saturation below 50% (see Table 14.1). The low cation exchange capacity and base saturation of Acrisols is the result of a long period of soil development in a humid warm climate. Because of the deep weathering and leaching of bases, the soils are strongly acid. Low activity clays such as kaolinite dominate the clay fraction. Illite and other 2:1 clay minerals can also be present but in small quantities. *Alisols* (formerly included with Acrisols) have a higher content of high activity clay minerals, including chlorite, smectite and vermiculite and a CEC of $> 24\,cmol(+)\,kg^{-1}$. They also have a high aluminum content and low base saturation and are, therefore, very acid.

Both Acrisols and Alisols contain finely distributed iron oxide and hydroxide, which form up to 10% of the fine grained soil, aluminum hydroxide, (gibbsite) is also often present. The silicate content in silt is low and the sand fraction is dominated by quartz. Silicate minerals that can be weathered are largely absent. The humus content is low forming either an ochric A horizon or an umbric A horizon if the base saturation is low.

Acrisols are nutrient poor and arable farming is possible only with the regular use of fertilizer which, with careful cultivation methods brings high yields Otherwise only shifting cultivation is possible, although with shorter fallow periods than on Ferralsols in the tropics, if there is a sufficient residue of minerals in the soil. The usable field capacity is more favorable than on tropical Ferralsols, but the Acrisols are disadvantaged because of a greater tendency to soil erosion than in the tropics. Phosphate fixation and aluminum toxicity also reduce fertility.

12.5
Vegetation

12.5.1
Structural characteristics

Where precipitation is high in the coastal regions and on the slopes of uplands, the natural vegetation is a dense *rainforest*. To the west inland, as the annual precipitation decreases, there are *semi-evergreen* moist forests or *evergreen* laurel forests, succeeded first by forest with deciduous species and then vegetation increasingly adapted to arid conditions. The tallest trees are between 20 and 25 m. In some regions there are plains of tall grasses, most extensively in the *pampas* in South America.

The Subtropical rain forests, although similar in appearance, have fewer species than tropical forests. Tree ferns, epiphytic ferns and lianas are common. In a rainforest near Brisbane, Australia, 2% of the basal area (of truncs) and 5% of the standing biomass were lianas (Hegarty 1991). The proportion of liana leaf mass compared to the tree leaf mass was even higher and consequently also its contribution to the litter (Table 12.1).

The laurel forests tend to have fewer species than the rainforests near the coast and the trees are usually lower in height with, at most, a two level canopy. Among the evergreen species are large numbers of laurels with sclerophyllous leaves although less xeromorphic than in the Subtropics with winter rain (Kira 1995). The leaves are large, oval and shiny. The proportion of deciduous trees in the zone increases westwards and towards the poles.

The overall pattern of east-west change in growth form distribution in the natural vegetation is not easily descerned. Destruction of the natural cover in the Northern Hemisphere in particular has progressed so far that it is difficult to reconstruct the original characteristic vegetation. The flora in all parts of the

Table 12.1. Structure and turnover characteristics in a subtropical rain forest near Brisbane, Australia

Life-form	No. of species	Basal area of trunks $(m^2 ha^{-1})$	No. of trunks >0.1 m height (ha^{-1})	Standing biomass $(t ha^{-1})$	Leaf mass $(t ha^{-1})$	Leaf fall $(t ha^{-1} a^{-1})$
Trees and shrubs	100	68.05	10,565	426	6.89	4.74
Lianas	42	1.56	5,771	21	2.52	1.47
Total	142	69.61	16,336	447	9.41	6.21

Because of its productivity in the canopy, the proportion of the leaf mass and the annual leaf drop of the lianas is much greater than its proportion of the trunk cross section area and of the biomass.

Source: Hegarty 1991

ecozone varies greatly so that it is probable that the natural plant formations were also greatly differentiated from one region to another in the past. The four areas of the ecozone in the Southern Hemisphere belong to three floral kingdoms, Neotropis in South America, the Palaeotropis in Africa and SE-Asia, and the Australis in Australia. New Zealand has its position between the palaeotropical and the antarctic kingdom. Both areas of the ecozone in the Northern Hemisphere belong to the holartic kingdom but lie far apart and have not been connected since the Pliocene

12.5.2
Available supply and turnover in a semi-evergreen oak forest in the southeastern United States

A forest in the southern Appalachians in the Coweeta Basin in North Carolina was examined by Monk and Day (1988). Their findings, based on observations made in the 1970's, have been largely confirmed by other researchers. The area lies on the borders of the Temperate midlatitudes. The stands are dominated by deciduous oaks. Evergreen trees and shrubs form 20 to 35% of the forest, based on the proportion of leaf dry matter. Rhododendron maximum, Kalmia latifolia, Tsuga canadensis, Pinus rigida, Ilex opaca and Leucothoe auxillaris var. editorum are the most important of the evergreen species. The first two, the rhododendron and kalmia, produce almost a third of the total leaf mass of the whole the forest, although their share of the total leaf area in the forest, with a mean of $6.2\,m^2$ per square meter of forest floor, is much less. This is because the *specific leaf area* (leaf area per leaf dry weight) of the evergreen is only $75\,cm^2\,g^{-1}$, half that of the deciduous.

The contribution of the evergreen leaves to the mineral supply of the total leaf mass of the forest is lower than their contribution to the dry matter because, compared to the deciduous foliage, their mineral contents are lower. The low requirement for the annual leaf production together with their long life is the reason for the dominance of evergreens on the mineral poor soils. In general, the distribution of large stands of evergreen is related both to the climate and the presence of mineral poor soils in any region.

The above ground biomass in the Coweeta Basin is small, $139.9\,t\,ha^{-1}$, but the below ground biomass is relatively large at $51.4\,t\,ha^{-1}$ (Fig. 12.5). Much of the forest was removed at the beginning of the 20th century and this coupled with the disappearance of the formerly dominant chestnut during the 1930's may be the reason for the low biomass values. The annual net primary production is $14.6\,t\,ha^{-1}$ of which $8.6\,t$ are above ground. From the NPP a total of $4.4\,t\,ha^{-1}$ is returned by litter fall and $0.2\,t\,ha^{-1}$ fed by herbivores. The remaining $4\,t\,ha^{-1}$ represents the increment in the stand, indicating that the forest is at a young or early mature stage (Fig. 5.3).

The difference between the leaf mass of $5.6\,t\,ha^{-1}$ and the leaf production of $4.2\,t\,ha^{-1}\,a^{-1}$ and between leaf production and leaf drop of $2.8\,t\,ha^{-1}\,a^{-1}$, is $1.4\,t$ in both cases, reflecting the importance of evergreen trees and shrubs. Leaf drop is also only half the leaf mass.

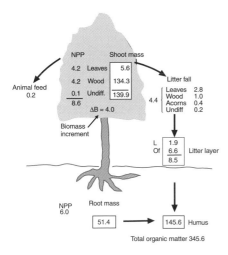

Fig. 12.5. Available supply $(t\,ha^{-1})$ and turnover $(t\,ha^{-1}\,a^{-1})$ of organic matter in a semi-evergreen oak forest in the Coweeta Basin in North Carolina U.S.A. (ΔB = increment) Source: Monk and Day 1988

The litter layer on the forest floor, the L and Of horizons, reaches $8.5\,t\,ha^{-1}$. With a litter fall of $4.4\,t\,ha^{-1}\,a^{-1}$, the duration of the decomposition is almost two years, a decomposition rate of 52%. Leaves compose 64% of the litter fall.

The quantities of minerals bound in the biomass are above those available in the soil, either in solution or exchangeable, for potassium, calcium and phosphorus. Only the supplies of magnesium and nitrogen are greater in the soil (Fig. 12.6). The high proportion of minerals in the vegetation in relation to the total of minerals in circulation is characteristic for many subtropical rainforests and moist forests.

The requirements for minerals in net primary production are supplied to a great extent from the leaves by resorption before leaf drop (Chap. 5, box 5). The differences in minerals between mature and old leaves, shortly before or after the fall, less loss from leaching, permit to estimate the resorption in $kg\,ha^{-1}\,a^{-1}$ as 56.5 for nitrogen, 13.5 for potassium, 3.4 for magnesium and 3.1 for phosphorus. Calcium is not resorbed (−0.5). The contributions of the resorbed minerals to the NPP are 54% for nitrogen, 26% for phosphates, 25% for magnesium and 21% for potassium.

The mineral content of the annual biomass increment in the stand can be calculated from the difference between mineral uptake and delivery during the same period. This indicates that only 6 to 9% of the nitrogen, magnesium, calcium and potassium uptake from the soil go into the remaining growth, only in the case of phosphorus is it 25% (the actual figures are N = 6.7, Mg = 0.9, Ca = 4.5, K = 4.9 and P = $3.0\,kg\,ha^{-1}\,a^{-1}$). This indicates that mineral consumption for the production of relatively short lived leaves is relatively wasteful compared with the mineral consumption for wood growth.

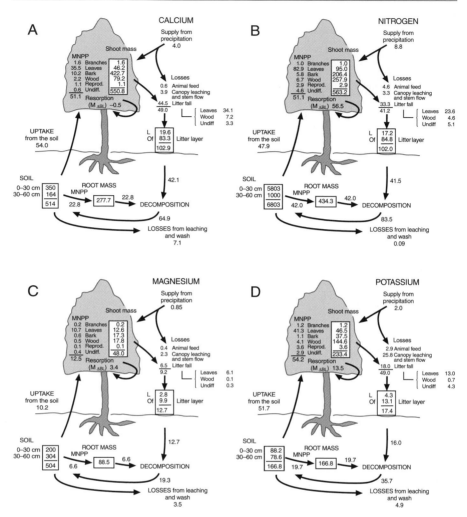

Fig. 12.6. Mineral nutrient budgets $(kg\,ha^{-1})$ and turnovers $(kg\,ha^{-1}\,a^{-1})$ in a semi-evergreen forest in Coweeta Basin, North Carolina, U.S.A. Data for mineral nutrients in the soil relate to the nutrients that are most easily available, either in solution or adsorbed by exchangers; Only in the case of nitrogen is the built in supply in the organic soil matter included.
Source: Monk and Day 1988

The minerals required for the biomass increment are equal to the loss in the available reserves in the soil. If a constant growth rate is assumed, it can be estimated how long a particular supply level in the soil will last. In the case described, calcium would be available for 138 years, potassium for 95 years, magnesium for 517 years, phosphorus for 12.5 years and nitrogen for 1,031 years.

A considerable proportion of the minerals are returned to the soil by canopy leaching, drops and stem flow. More K^+ returns to the soil by this means than

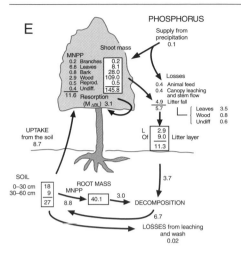

from the litter. This is true to a lesser extent for SO_4^{--}, PO_4^{---}, Cl^-, Ca^{++} and Mg^{++}. The annual return of minerals to the soil from the litter and by leaching are sufficient, with the exception of phosphorus, for more than 8 years wood growth in the stand at a constant rate.

Nitrogen is an exception to other minerals because the supply from precipitation is much higher than the output as a result of canopy leaching so that there is a net gain by absorption in the canopy.

12.6
Land use

Most regions of the Subtropics with year-round rain are densely settled and economically highly developed. The natural vegetation has been largely removed and replaced by a cultural landscape of settlements, large industrial complexes, land for agriculture and for forestry. (Fig. 9.11).

In summer, high temperatures allow the cultivation of plants requiring warmth. Since sufficient precipitation falls, no additional moisture is required. Winters are also mild with only occasional light frosts. This allows the cultivation of tree crops, such as citrus and tea are also grown. Severe winters may, however, cause extensive damage to such crops. Among the annual crops are sorghum, maize, peanuts, rice, soya, sesame, sweet potato, cotton and tobacco. Plants from the midlatitudes are cultivated in winter so that two or even three harvests are possible annually.

Where land use is the result of settlement by Europeans, farms are generally middle-sized and specialize on one crop or animal products for the market. Labor requirements are low, farms highly mechanized and productivity per unit and man hour high. Although the soils are not naturally productive, this is not a limitation if fertilizers are applied and modern agricultural methods used. Only in southeast China is traditional rice cultivation in small holdings still dominant.

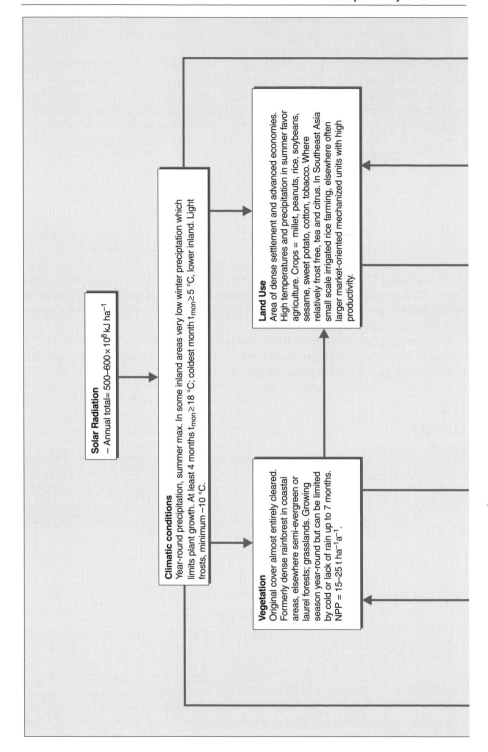

Solar Radiation
– Annual total= $500–600 \times 10^8$ kJ ha^{-1}

Climatic conditions
Year-round precipitation, summer max. In some inland areas very low winter precipitation which limits plant growth. At least 4 months $t_{mon} \geq 18$ °C; coldest month $t_{mon} \geq 5$ °C, lower inland. Light frosts, minimum −10 °C.

Vegetation
Original cover almost entirely cleared. Formerly dense rainforest in coastal areas, elsewhere semi-evergreen or laurel forests; grasslands. Growing season year-round but can be limited by cold or lack of rain up to 7 months. NPP = 15–25 t ha^{-1}a^{-1}.

Land Use
Area of dense settlement and advanced economies. High temperatures and precipitation in summer favor agriculture. Crops = millet, peanuts, rice, soybeans, sesame, sweet potato, cotton, tobacco. Where relatively frost free, tea and citrus. In Southeast Asia small scale irrigated rice farming, elsewhere often larger market-oriented mechanized units with high productivity.

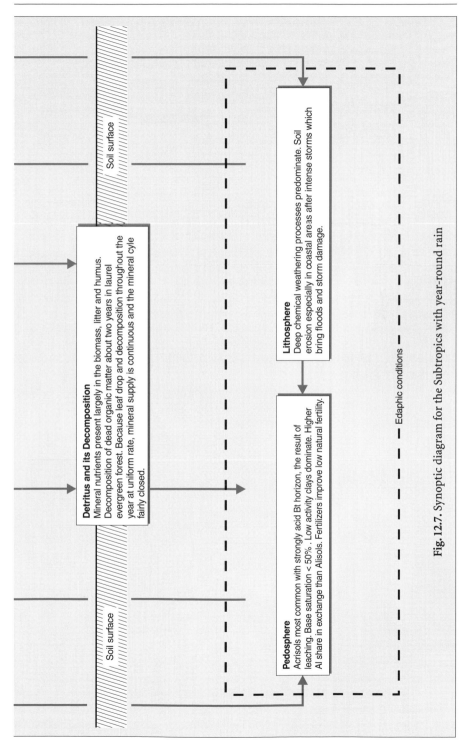

Detritus and its Decomposition
Mineral nutrients present largely in the biomass, litter and humus.
Decomposition of dead organic matter about two years in laurel
evergreen forest. Because leaf drop and decomposition throughout the
year at uniform rate, mineral supply is continuous and the mineral cyle
fairly closed.

Soil surface

Lithosphere
Deep chemical weathering processes predominate. Soil
erosion especially in coastal areas after intense storms which
bring floods and storm damage.

Pedosphere
Acrisols most common with strongly acid Bt horizon, the result of
leaching. Base saturation < 50% . Low activity clays dominate. Higher
Al share in exchange than Alisols. Fertilizers improve low natural fertility.

Edaphic conditions

Fig. 12.7. Synoptic diagram for the Subtropics with year-round rain

CHAPTER 13

Dry tropics and subtropics

13.1
Distribution

The Dry tropics and subtropics ecozone includes not only semi-deserts and deserts but also two semi-arid transition zones to more humid neighboring ecozones: the wet summer *thorn savanna* and the *thorn steppe* which lie in the transition zones to the Tropics with summer rain and the Subtropics with year-round rain, and the *grass and shrub steppes* with winter rain which are in the transition zone to the Subtropics with winter rain (Fig. 13.1). The total area of the ecozone is 31 million km^2 or 20.8% of the world landmass.

Both the boundaries of the zone and the divisions within it are closely related to the distribution of the mean annual precipitation (Table 13.1). The lower threshold values for precipitation in the border areas towards the poles are an expression of the lower air temperatures and lower transpiration burden for plants in these regions.

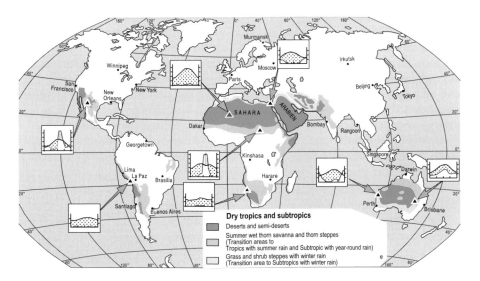

Fig. 13.1. Dry tropics and subtropics

Table 13.1. The boundaries of the subdivisions of the Dry tropics and subtropics related to mean annual precipitation (cf Fig. 13.5)

	Boundary between	Approximate annual precipitation (mm)
Equatorwards	Desert and semi-desert	125
	Semi-desert and thorn savanna	250
	Thorn savanna and dry savanna in tropics with summer rain	500
Polewards	Desert and semi-desert	100
	Semi-desert and steppes with winter rain	200
	Steppes with winter rain and sclerophyllous shrubs in Subtropics with winter rain	300

13.2
Climate

Almost all warm arid regions lie wholly or partially within the subtropical/outer-tropical high pressure belts of the Northern and Southern Hemispheres. Characteristic of the high pressure cells is a constant subsidence so that the air at the surface is warm and dry and the air layers stable to a considerable height. Thermal convection leads only rarely to cloud formation or precipitation. Annual cloud cover is therefore low, usually less than 30%, in some areas less than 20%. The Dry tropics and subtropics receive more global radiation annually than any other zone, including those at a similar latitude or closer to the equator.

A large proportion of the solar radiation is reflected immediately. The arid regions have much a higher albedo than areas covered with vegetation. Although here too, there can be a considerable range in the albedo, depending on the texture, color and moisture content of the soil and the type of vegetation. In deserts the albedo generally lies between 25 and 30% so that the amount of energy received is relatively small compared to the high global radiation.

If during the day the earth's surface (soil, rocks and plants), should be subject to an extreme heating up, this can be attributed to the prevailing dry substrate, where

(a) the absorbed radiation energy (net incoming short-wave radiation) is transformed almost completely to tangible warmth (hardly any latent heat flows as is usually the case with evaporation from moist substrates: Fig. 13.2) and

(b) the heat conductivity and capacity of the soil layer (due to many air filled pores) are very low, so that the energy intake is concentrated in the uppermost centimeters.

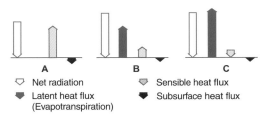

▽ Net radiation   ▽ Sensible heat flux
▼ Latent heat flux   ▼ Subsurface heat flux
  (Evapotranspiration)

Fig. 13.2. Radiation and energy balance at noon in A.a desert, B a moist surface C. an oasis. In the desert almost all net radiation is balanced by the sensible heat flux and, as a result, the rise of soil surface and air temperature is high during day-time. The latent heat flux of a moist surface and oasis are higher because of the availability of soil water. In an oasis the total of latent and sensible heat fluxes can surpass the net radiation received because warm dry and hot air flows from the surrounding desert area and brings heat and therefore additional energy for evaporation. This is known as an oasis effect and can be present as well in narrow humid areas on either side of rivers in dry areas and is also the exploration for the often much greater than estimated evapotranspiration losses in the immediate vicinity of reservoir and irrigation projects in dry areas. Source: Rouse 1981

On the other hand high surface temperatures increase the outgoing long-wave radiation. In addition to this, the atmospheric counter radiation remains extremely small due to the low moisture level in the air (steam is by far the most important green house gas in the atmosphere), so that the effective loss by thermal radiation is exceptionally high. The consequence is a very negative radiation balance at night(Chap. 2 Box 1), as a result of which temperatures sink rapidly. The diurnal amplitude is therefore extremely high, with the consequence that rock disintegration processes by thermal expansion and contraction play a major role.

The lack of cloud cover and low humidity means that up to 75 % of the annual radiation is in the form of *direct radiation*, and even more in the dry season. Consequently, heating up of the earth's surface and evaporation vary considerably on the differently exposed slopes. This may be reflected by the distribution pattern of vegetation and rock debris (in particular of thermal weathering).

Mean annual precipitation is highly variable from year to year and from place to place in all arid regions. Plant and animal life is often under stress. Hot, dry ecosystems can be defined as moisture dependent ecosystems. All other environmental factors, air temperature, radiation and usually also mineral nutrient supply in soils, are favorable, or at least not limiting, except in some upland areas or on dunes.

In order to be effective for vegetation growth, a precipitation event must total at least 10 mm. If less, the moisture is evaporated before it can be absorbed by the roots.

13.3
Relief and drainage

13.3.1
Weathering processes and crusts

Despite the dry conditions, *chemical weathering* also takes place in arid areas as indicated by the presence of salt, in many soils. In general, however, it is much less important than the mechanical weathering, except, e. g. on level surfaces or valley bottoms, where deep regoliths could develop. Since the products of chemical weathering are not removed or leached out, capillary movement in the soil leads to enrichment in soil layers near the surface and hardening at the surface. The *crusts* that then form range from easily soluble salts, usually sodium chloride, to duricrusts of calcrete, a secondary precipitate of $CaCO_3$, gypcrete a precipitate from $CaSO_4$ or silcrete from SiO_2.

Mechanical weathering in arid regions is generally either in the form of *salt shattering* or of *thermal weathering*. Salt shattering takes place where rain or dew penetrates the pores or fine cracks in the rock and releases salts and other materials formed earlier during hydrolysis. In dry periods capillary movement brings salts to the surface layer where they crystallize out. The pressure resulting from the crystallization process is increased when the growing crystals are hydrated. Repeated soaking and drying out increases the pressure further which in turn widens cracks and loosens mineral grains in the rock to produce a shattering effect. Capillary movement may also bring iron and manganese to a rock surface which then acquires a dark coloring known as desert varnish.

In thermal weathering, expansion and contraction that create a tension in the rock are caused by temperature changes. The resulting volume changes in the rock lead to granular disintegration, thermal exfoliation, flaking and block disintegration. Chemical weathering may also play a role in thermal weathering processes, as do other physical processes.

Angular fragments from the size of sand grains to blocks accumulate around the weathered rock and because there is little erosion or wash denudation in arid areas, thick layers of debris remain both on the slopes of uplands and at the slope foot. In rock deserts such as hamadas, the lower rocky slopes may disappear under debris.

13.3.2
Aeolian processes

Aeolian processes are important in arid areas in which there is no vegetation and the effectiveness of flowing water is limited. The forms they produce are most frequent in semi-deserts and deserts, although even in these environments they do not cover large areas.

The wind transports sand grains either by saltation, a process by which the grain is lifted from the ground, rarely more than one meter in height, and moved in stages in a low curve with the wind, or by reptation whereby

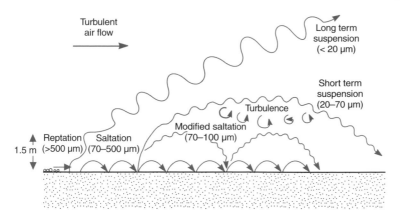

Fig. 13.3. Types of wind transport. The diameter of the particles moved is shown for each type of movement. Source: Pye 1987

the impacting of springing grains moves other grains which leads to a form of surface creep (Fig. 13.3). The ratio of sand moved by saltation to reptation is 3:1.

Dust is not moved by saltation but held in suspension by turbulence and carried to great heights and moved long distances.

Deflation, wind abrasion and accumulation are the geomorphological effects of wind. Deflation is the blowing out of loose material by the wind over large areas from an originally unsorted regolith. Extensive areas of desert pavement result, or, where there were deep layers of fine grained material, deflation hollows. Deflation can also cause narrow cuts in dunes. The forms in gravel deserts (serir or reg) are probably at least partially formed by deflation.

Sand transported by wind has an abrasive effect on surfaces over which it passes. The areas affected are below one meter in height, the maximum height of saltation, or at most two meters during sandstorms. Along the foot of rock faces, abrasion forms hollows which develop into individual mushroom like forms if the wind blows from several directions. If the wind blows from only one direction, deflation develops elongated furrows separated by long streamlined ridges known as yardangs. Individual stones on the ground can also be faceted by abrasion on their windward side.

Dunes are formed when sand is redeposited. In some areas there are large complexes of dunes in sand deserts that cover more than a thousand square kilometers, although in total they account for only a few percent of the desert regions.

13.3.3
Stream erosion and wash denudation

Although discharge events are of short duration, the movement of material by flowing water has a greater impact than aeolian processes in even the dryest areas. Drainage in arid areas, apart from exotic streams, is episodic. The

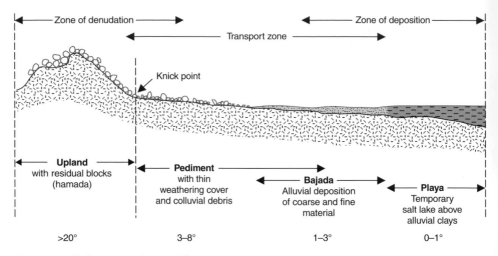

Fig. 13.4. Relief sequence in an arid area

streams are supplied by surface or near surface flows of water, usually related to a particular precipitation event, and often soon disappear again underground where flow may continue for some distance. Large quantities of sand and pebbles can be moved during a precipitation event because of the high velocity of the water. The slopes of the uplands often become deeply incised by streams where they border the plain. Gently inclined pediments develop in this border area to the uplands. They are levelled by wash denudation and lateral erosion as the stream exits from the uplands. The streams also lose gradient as they leave the upland areas and deposit their loads in large alluvial fans or alluvial cones composed of coarse fragments and sands that have hardly been rounded by the action of the water. Beyond the pediments on the plains are broad shallow valley bottoms or sedimentary basins into which the streams flow.

13.4
Soils

According to the world soil map of FAO–UNESCO soils in the most extreme arid areas are *Yermosols* and in the semi-arid regions mostly *Xerosols*. They are defined as aridic with a very weak or weak ochric A horizon, respectively. If the ochric horizon is very weak, the humus content is at most 1% of weight in the upper 40 cm in clays and less than 0.5% in weight in sands. Xerosols which have a sparse but denser plant cover than Yermosols also have a slightly higher humus content.

Yermosols and Xerosols are no longer part of the new FAO soil classifications. The most frequent soils in the arid and semi-arid regions under the new classifications are termed Calcisols, Gypsisols, Arenosols and Regosols etc. (Chap. 4, Box 2).

Arenosols are coarse textured sandy soils with a very low fine grain content of aeolian or marine origin or from in situ weathering of rocks with a high quartz content. The soil profile is hardly differentiated. An ochric A horizon may have a weakly developed B horizon below.

Calcisols, Gypsisols and *Durisols* are characterized by calcium carbonate, calcium sulphate and silicium oxide (with secondary quartz and opal) enrichment in the subsoil. The enrichment develops as the water percolates from the surface carrying material in solution. The consistency of the soils ranges from powdery to hard cemented *duricursts*. All three soils are found mainly in the semi-arid transition zones. Soils in the thorn savanna and thorn shrub steppe have very weak humus development and there is no equivalent of the humus rich soils in the steppes of the Dry midlatitudes.

13.5
Vegetation and animals

About three-fifths of the Dry tropics and subtropics are deserts or semi-deserts. The vegetation in the transition areas on the borders of the zone vary according to whether they receive precipitation in summer or in winter. Where there is winter rain, on the borders of the Subtropics with winter rain, *grass and shrub steppes* dominate. In the areas where there are peaks of summer rains, the predominate vegetation is *thorn steppe* and thorn *savanna*. The tropical thorn savanna lies in the transition zone to the savannas of the Tropics with summer rain. Most of the subtropical thorn steppes lie in the eastern part of the Dry tropics and subtropics in the transition area to the Subtropics with year-round rain, from where, towards the equator, the thorn steppes merge into the thorn savanna (Figs. 13.5 and 13.1).

Table 13.2 shows the coverage of perennial vegetation and distribution of trees in the dry regions of the tropics and subtropics, compared to savannas in the Tropics with summer rain.

Semi-deserts are areas with an upper limit of grass cover of 50%. Trees are confined, for the most part, to dry valleys and the foot of upland areas Herbaceous plants and dwarf shrubs are evenly distributed. The transition to *deserts* lies where growth of perennial plants becomes restricted to a few most favored patches which altogether do not cover more then 10% of the area (Fig. 13.7). In the extreme deserts, there is no perennial vegetation at all, at least not of vascular plants.

Where grass cover is over 50%, although still with large gaps in the cover, thorn and shrub steppes or thorn savanna has developed. Typically, instead of the linear growth pattern of semi-deserts, trees cluster in hollows and, where conditions become more favorable, in an even growth which increases

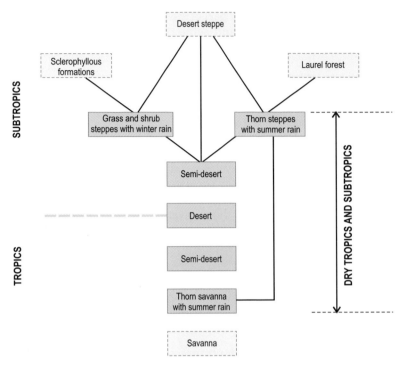

Fig. 13.5. Vegetation in the Dry tropics and subtropics in relation to the neighboring zones. Broad transition areas often exist between the plant formations of the zones.

in density as the moisture supply increases until eventually the trees become the dominating element in the landscape.

13.5.1
Vegetation and soil water budget

The availability of water can range widely, both regionally and locally, from the climatic water budget and underlies the wide variation in the vegetation patterns in the ecozone.

Since vegetation is sparse at the most, impacting rain drops, especially heavy drops, strike the soil and cause particles of sand, silt and clay as well as soil aggregates to move in all directions. The result of this splash process is a partial sealing of the surface with fine material and a reduction in the *infiltration capacity* of the soil. In arid areas and also areas that are seasonally dry, a high proportion of the precipitation does not infiltrate where it falls but, even on gently inclined slopes, flows over the surface into playas, dry valleys or over pediments at the base of the uplands (Fig. 4.3).

The proportion of the precipitation that does not infiltrate, depends on the intensity of the precipitation events, the inclination of the surface, the

Table 13.2. Characteristics of the vegetation in the Dry tropics and subtropics and the savanna of the Tropics with summer rain

Characteristics	Desert	Semi-desert	Thorn savanna Thorn steppe Shrub steppe	Savanna of the Tropics with summer rain
Vegetation cover (%)	usually < 10	10–50	> 50 with gaps	100
Distribution of perennial grasses, herbs and dwarf shrubs	patchy	———— diffuse ————		closed
% chamaeophytes in the herb layer	———————— > 50 ————————			approaching zero
Therophytes	large number of species ◄——————————► few species			
Distribution of trees	linear (in dry valleys and foot of upland areas)		clusters to widely distributed	
Growth height – herb layer – tree layer	◄——— < 50 cm ———► ◄——— < 5 m ———►		< 80 cm 5–10 m	80 up to > 200 cm 10–20 m
Biomass of herb layer	extremely low	very low	max. 2–5 t ha⁻¹	usually > 5 t ha⁻¹
Root/shoot ratio	2–5 ◄———————————————————► ≈ 1			
Above ground NPP of herb layer – per mm of annual precip. (kg ha⁻¹ a⁻¹)[a] – total (t ha⁻¹ a⁻¹)	0–1 ——— usually < 1 ———	1–3	4 1–2.5	5–7.5 > 2.5

[a] Rain use efficiency

vegetation cover, the surface roughness and the texture and depth of the soil and subsoil. The surface runoff is high when the regolith has a low infiltration rate because of a fine-grained loamy or clayey texture, the bedrock is bare, the slopes steep, the area covered by vegetation small and the rainfall heavy.

Although aridity is increased in one area by surface runoff, conditions elsewhere are created that can support more growth than would be possible if only the local rainfall were available. The *spatial concentration* of water in the desert is of advantage to plants and animals because at least temporarally, in some areas and on some soils, conditions are suitable for plants to grow and animal life to survive.

The soil texture does not only influence the infiltration rate, but also determines the quantity of infiltrated water that can be stored in the soil. If the soil texture is fine and the field capacity therefore high, the infiltration depth and volume of wetted soil are small. Also a relatively large share of the bound

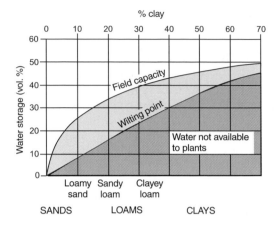

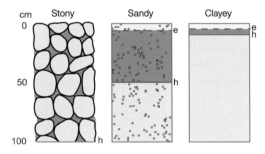

h = Lower boundary of soil wetting

e = Lower boundary to which soil dries out again

Fig. 13.6. Water storage capacity of soils with different grain sizes. The maximum field capacity increases with decreasing grain size. In clayey soils the FC is therefore greater than in sandy soils. An assumed rainfall of 50 mm percolates 100 cm into stony, 50 cm into sandy and 10 cm into clayey soils. Source: Achtnich and Lüken 1986, Walter and Breckle 1983

water can be lost following capillary movement and evaporation or is not available to plants (Fig. 4.4). With a coarser texture and lower field capacity, water infiltrates more deeply. All the water is usable because it is only loosely bound to the soil particles and following the drying out of the upper soil layers is protected from further evaporation. There is also no capillary movement (Fig. 13.6). In dry areas, soils on debris or sands have, under equal conditions, a more favorable water budget than soils in clayey material, providing roots can penetrate deeply enough.

Large amounts of water that become available locally from lateral run off are not always advantageous because they bring easily soluble salts with them. Many plant species are sensitive to salt in the soil. Even salt tolerant plants may suffer if salt concentrations become very high. Depending on the degree of salt stress, the vegetation cover is differentiated in a mosaic pattern or in strips around salt pans and along dry stream beds.

Of significance to the water supply of an individual plant is its distance from the neighboring plants because this determines the volume of water that is available to the roots and how much they can take up. It is possible for a precipitation deficit to be compensated for by a more developed root system. There is, therefore, an optimal distance between plants related to the availability of moisture locally. In arid areas a plant cover in which the plants are dispersed has a much denser root development in the soil because the roots compete for the water supply, unlike more humid areas where they compete for light.

As mean annual precipitation declines below 100–125 mm, the available water is no longer sufficient for root development to compensate for the deficit and vegetation growth is only possible in small, frequently linear areas where soil moisture is above average for example by lateral inflow (Fig. 13.7).

Above an annual precipitation of approximately 125 mm the vegetation ranges, depending on the moisture conditions, from a widely dispersed cover as found in semi-deserts to a relatively closed cover as found in the steppes and savannas. Grasses predominate on clayey soils and trees, with their extensive root systems, in sandy conditions.

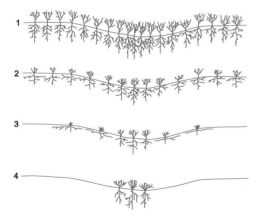

Fig. 13.7. Transition from diffuse vegetation cover to patchy vegetation in relation to declining precipitation in a very dry area. Compared to the moister thorn savanna and steppes, plant cover in a semi-desert is patchy and the roots have a larger area from which to obtain water. Patchy plant growth is limited to areas in a desert, such as a wadi, where water accumulates at the surface or subsurface and root depth is related to the local depth of the water supply or groundwater. Source Walter and Breckle 1983

13.5.2
Adaptation to drought and stress

Plants have developed different methods to survive *drought* and *salt stress* (Fig. 13.8). Annual plants adapt to drought by developing rapidly, at least above the ground. Perennial herbaceous and woody plants have the ability to slow the drying out and also to store water. *Halophytes*, in arid environments, adapt to both drought and high salt concentrations.

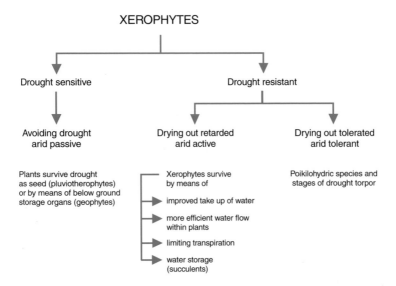

Fig. 13.8. Means of survival of plants in dry regions. Source: Larcher 1994

In general the share of woody plants, mostly low shrubs, semi-ligneous shrubs and trees, but also annuals (therophytes) increases with declining precipitation. Woody plants are, therefore, more numerous in semi-deserts and deserts than in grass steppes and thorn savanna where perennials (hemicryptophytes), mostly members of the grass family (Gramineae), dominate.

Figure 13.9 shows typical woody growth forms. Despite the constant supply of solar energy, it is only for short periods after rainfall that plant growth is possible. Figure 13.10 shows growth patterns of ephemeral herbs, perennial grasses and evergreen woody plants in arid regions after a precipitation event.

Plants vary greatly in the way they react to arid conditions. The water budget of plants that tolerate drying out, adjust to the moisture conditions of their surroundings. With continuing loss of moisture, the plant goes into a drought torpor until the next soaking. A large number of low plants are drought tolerant, including fungi, algae, lichen and blue-green algae, but also some vascular plants. Normally only the spores, seeds or pollen are drought tolerant.

Other xerophytic plants are able to improve their intake of water as well as slowing down the losses by transpiration. In this type of plant, the water

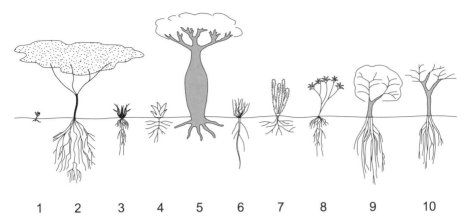

Fig. 13.9. Characteristic growth forms in the thorn succulent savanna. 1. Pluviotherophytes, 2. Thorny fine pinnate umbrella trees or shrubs (Acacia), 3. Hard grasses with buds surrounded by a sheath and deep roots (Aristida), 4. Succulent leaves (Agava/Crassulaccac), 5. Deciduous trees with broad trunks and water retaining trunks (Adansonia/Chorisa), 6. Switch shrub (Retama), 7. Stem succulent (Cactus/Euphorbia), 8. Tufted trees with succulent leaves, 9. Evergreen trees and shrubs with deep roots (sclerophyllous) and thorns (Balanites), 10. Deciduous often thorny trees and shrubs (Commiphora). Source Larcher et al. 1994

intake can take place largely over a highly developed root system and a reduced shoot surface on which transpiration takes place. Roots may be very densely developed in a limited soil volume, or they can form an extensive system reaching to a depth of 8 to 10 meters or, spread as rain roots just under the surface but extending around the plant over a large area. Many xerophytics develop a higher root suction pressure by concentrating the fluid in their cells up to a pressure of 60 atm or more compared to 10 to 20 atm for most mesophytes.

Other characteristics in various species of xerophytes are pinnate leaves, a reduction of the assimilating organs to thorns, leaves that fall during the dry season and grasses that die off above the surface. Some species roll their leaves along the central vein during periods of stress from drought or diurnally, to avoid the sun's rays.

Low growth is common among xerophytic plants, because limited transpiration results in reduced uptake of minerals from the soil and also reduced exchange of gases during photosynthesis. As aridity increases, the heights of plants decrease.

Succulent xerophytes combine the ability to delay drying out in the plant with water stored during moist periods. Succulents are classified according to whether they store water in leaves, stems, shoots or roots. Shoots that are succulent also have a smaller surface for transpiration in relation to their biomass, which increases their ability both to store water and to protect against water loss. These advantages are offset, however, by a reduction in their light absorbing surfaces so that there are fewer cells for photosynthesis.

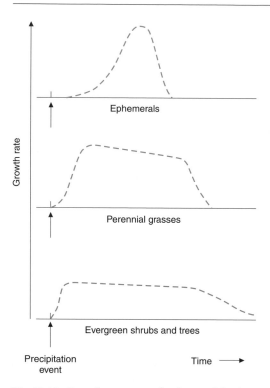

Fig. 13.10. Growth patterns of ephemeral herbs, perennial grasses and evergreen woody plants in arid regions after a precipitation event. Ephemeral herbs have the highest photosynthesis capacity and the evergreen woody plants the lowest. The growth curve of the ephemerals increases last in response to the precipitation event because at first seeds have to germinate or shoots from the subsurface organs to grow and leaves to develop for assimilation. The growth curve of ephemerals also ceases earlier because ephemerals use the soil water more rapidly. The lower the rate of transpiration of the perennials, the more moisture remains in the soil and the longer are the growth periods. Source: Ludwig et al. 1997

Many succulents have developed thorns instead of leaves which helps to reduce the plant surface, prevents eating by animals and protects from heating by the sun. Photosynthesis in cactus and many other succulents takes place using the CO_2 stored in the plant during the previous night. In this way the stomata can remain closed during the day.

Some xerophytes avoid drought by confining their budding and germination to the period beginning with the onset of rains. Their entire development including the ripening of the seeds occurs within a few months, somtimes in as little as one, and before the available moisture is used up. Many of these plants do not have xerophytic characteristics. In general, their adaptation is a functional rather than a structural adaptation to aridity.

Annual xerophytes have seeds that survive drought by going into a stage during which the seeds are dried out (pluviotherophytes); others have rhizomes, bulbs or tubers underground in which they can store water (geophytes). Because they appear only when it rains, these types of plants are termed ephemeral, in contrast to permanent vegetation. They are almost the only vegetation in areas of extreme aridity in which the mean annual precipitation is only 10 to 20 mm.

High concentrations of salt in the soil burden the plants during droughts in two ways. The intake of water is impeded because the water in salt solution is bound osmotically and because the plants have a disproportionately large intake of sodium and chlorine, as opposed to calcium and potassium, resulting in an ion disequilibrium which leads to a physiological impairment of the plant. Some halophytes combat an excessive intake of salts by lowering their own osmotic potential so that even with high concentrations in the soil solution, a potential gradient towards the roots remains. A negative effect on the protoplasm can be avoided because the plants store relatively large amounts of water in their shoots, this is described as salt succulence. A few plants, tamarisks for example, discharge excess salt.

13.5.3
Animals in the desert

The species diversity and also the population densities of animals in deserts are low and vary widely in relation to the spatial distribution of vegetation and its changes in time. The contribution of the consumers to the turnover in the ecosystem is temporarily and spatially very uneven and generally small. At

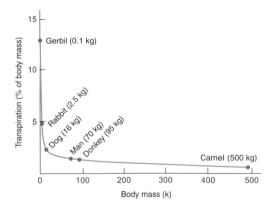

Fig. 13.11. Mammals use transpiration to regulate their body temperature during periods of heat stress. The quantity of moisture required depends on their heat uptake which increases with the ratio of surface to mass. Relative to body size, small animals must transpire more than larger animals. Gerbils (rodents) of 0.1 kg lose 13% of their body weight per hour, camels of 500 kg lose 0.8%. It is assumed that the heat flow per cm^2 of body surface is equal for all cases and therefore proportional to the animal's surface area. The heat interception per kg of the body mass is, therefore, in reverse proportion to its relation to the body surface. Source: Lovegrove 1993

most 2% of the biomass is returned as a result of feeding. The interaction between plants and animals is quantitatively relatively unimportant since major changes to one or other biotic component are of little consequence to the rest of the system.

In deserts, animals have to have the ability to reduce their output of water and regulate body temperature. Either they adapt their morphology and, or, physiology or they adapt their behavior. Behavioral adaptation to avoid stress includes retreating underground during the extreme heat of the day or cold of the night to where humidity is high and temperature moderate. Large mammals can deal with heat stress better than small animals because their heat intake in relation to their body mass is lower. The amount of transpiration necessary for cooling relative to their body weight is too large for small animals (Fig. 13.11). To avoid the hot arid conditions most invertebrates and small vertebrates are active only at night.

13.5.4
Biomass and primary production

Mean values for biomass and primary production are low in the ecozone because of the unfavorable growing conditions. There are considerable fluctuations in supply and turnover which lead to short-term instability. The ability of desert vegetation to react flexibly means that following a major precipitation event, primary production can increase rapidly and an above ground biomass develops that surpasses that of dry years several times. Ephemeral plants contribute 50% or more to that exceptional large biomass. Most of the arid active plants and most animals can synchronize their development phases to periods

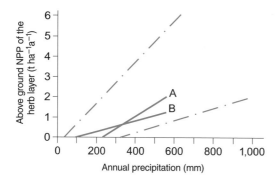

Fig. 13.12. Relationship between annual above ground production of the grass/herb layer, annual precipitation and soil fertility/texture. A shows rain use efficiency (RUE) on fertile clay soils with 5 kg dry matter $ha^{-1} a^{-1}$ per mm rainfall; B shows a RUE on sandy soil but with 2.5 kg dry matter $ha^{-1} a^{-1}$ per mm precipitation. The correlations can be used to estimate, for example feed yields in other regions of similiar climate or as a measure of fertility. The broken lines show the boundaries for RUE values in other studies. Source: Scholes 1990, Rutherford 1980

of the year when water is available. Based on this elasticity a long term stability develops. Desert ecosystems, therefore, cannot be regarded as more fragile to intervention by man than other ecosystems.

The annual primary production in the most arid deserts, ranges from about zero to almost $0.2\,t\,ha^{-1}$, in semi-arid regions it rises to $2.5 - 3.0\,t\,ha^{-1}$. Regional differences in above ground production are related primarily to variations in the amounts of precipitation, but can also be affected by soil fertility and other local factors (Fig. 13.12). The mean rain use efficiency of all arid areas has been estimated at a rain factor of 4, that is an annual above ground production of $4\,kg$ dry matter ha^{-1} per millimeter of annual precipitation, with a range of 1 to 10 (Le Houerou et al. 1988).

Rain use efficiency in a region can be used not only to estimate the mean annual production based on the mean annual precipitation, but also the production in any one year on the basis of the precipitation in that year.

13.6
Land use

The Dry tropics and subtropics lie beyond the arid limit for rainfed agriculture. The human carrying capacity in the ecozone is very small. Where there is cultivation outside of oases, e.g. in parts of the Sahel in Africa, it is largely with plants that require little water such as millet (pearl millet) and peanuts or rapidly maturing species, including some beans. The yield of all these crops is uncertain and the risk of soil erosion high because of wash denudation and deflation of fine material both of which remove organic material and mineral nutrients from the soils.

Extensive grazing and the cultivation of crops that can be irrigated are economically and ecologically more sustainable. For all forms of agriculture sustainable development has to take place within the restraints of the natural conditions, particularly the variability of the rainfall. Restoration of land surfaces destroyed or damaged by over grazing or natural causes takes much longer than in other environments and may not be possible at all. Regeneration is possible if protective measures are taken by restricting grazing or if precipitation is higher in one year.

13.6.1
Pastoral nomadism

In the arid regions of Africa and Asia grazing on natural pastures in the form of nomadism has long been important. In South America, southern Africa and Australia extensive grazing is practiced in a stationary (and commercial) way called ranching (cf. Chap. 10.6.2). At the present time in Asia and Africa most groups of formerly nomadic herdsmen are now semi-nomadic or practice transhumance. Year-round nomadism is seldom (Fig. 13.13).

Camels, sheep and goats form most herds in the desert, semi-desert and subtropical steppes. Cattle dominate in the thorn savanna. The difficult natural

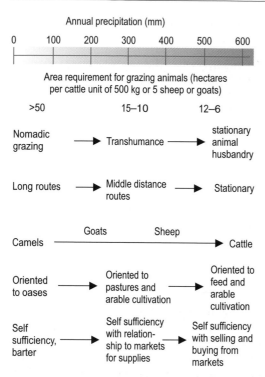

Fig. 13.13. Traditional pasturing economies in relation to moisture conditions (corresponds to the carrying capacity of natural pastures). Natural conditions determine the type of farming. The better the quality of the pasture, the shorter the distances covered and the closer the relationship to the markets. Source: Doppler 1991

conditions favor breeds that can survive in dry conditions when the quality of feed is poor and it is necessary to migrate long distances. The quality of meat and milk is low.

Nomadic grazing takes place beyond the limits for ranching. Without the nomads large areas of the arid regions would remain unused. The amount of labor required to look after herds is relatively high but the income is small. Many nomadic groups have also lost their function as transporters of goods for trade. In addition, in some areas arable farming has extended on to the grazing land formerly used by the nomads which also means the removal of the nomads' best pasture lands. The present limit of arable land in the thorn savanna lies where the annual precipitation is about 400 mm or even less, well below the boundary for more or less reliable crop yields.

The lower the productivity of the natural pastures, the higher the area required per animal, as annual precipitation declines these area requirements also increase (Table 13.3). An estimate of the amount of dry matter suitable for feed that has to be produced from pasture is generally put at $1 \, \text{kg ha}^{-1}$ per millimeter of annual precipitation, about 25% of the above ground net primary

Table 13.3. Area required per grazing animal on arid and semi-arid natural pastures in the tropics in relation to mean annual precipitation

Annual precipitation (mm)	Required area (ha) per livestock unit	Number of cattle per 100 hectares
50–100	> 50	< 2
200–400	15–10	7–10
400–600	12–6	8–17

Livestock unit = 1 cattle of 500 kg living weight or 5 sheep/goats

Source: Ruthenberg 1980

production. Animals usually need daily between 2.5 and 3.5 kg of dry matter per 100 kg of weight (Le Houerou 1989).

13.6.2
Oasis agriculture

Irrigated cultivation is the only reliable form of agriculture in arid areas because it is independent of rainfall variability and produces high yields from a wide variety of field and tree crops. Soils are often fertile and the amount of solar energy available is very high so that year-round use of the land and several harvests are possible. The natural high yield potential depends on sufficient non salty water being available. Water for irrigation comes from exotic rivers, the groundwater, storage of rain water or from desalination plants on the coast. Various techniques are used to control the water resource. They include large dams to store water for irrigation projects, the partial covering of rainwater flow on the surface; small dams or ditches on the fields to concentrate flow (Fig. 13.14), the transportion of groundwater considerable distances by pipeline and the drawing of water from local wells using pumps or water wheels.

① **Micro catchments** with semi-circular, funnel shaped or trapezoid retaining walls of earth or rocks around planting basins for a tree or small cultivated area for annual cultivation

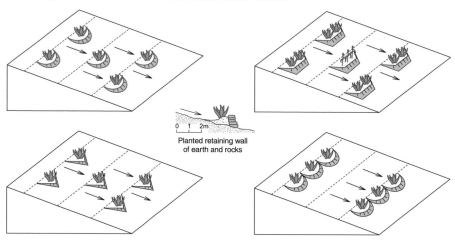

0 1 2m

Planted retaining wall
of earth and rocks

② **Contour strips** and **contour terraces**. In the simplest case there are furrows and dams parallel to the contours, levelling of the adjoining slope area extends the cultivated area

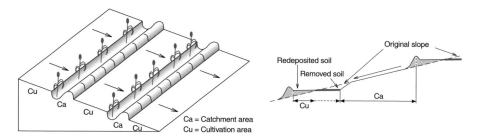

Ca = Catchment area
Cu = Cultivation area

③ **Within field catchments.** Creation of artificial catchments (low ridges) within the field area. To improve runoff the ridges can be made impervious or covered with plastic sheets.

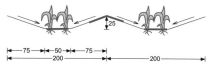

Typical size of catchments and cultivated areas in cm

Fig. 13.14. Examples of cultivation using concentrated water flows. The surface runoff from precipitation is lead by means of a natural (1 and 2) or artificial gradient (3) from a relatively large catchment area to a tree or small cultivated area. The relationship between the size of the cultivated area and the catchment area is a function of the degree of aridity. Source: Finkel 1986 UNEP 1983

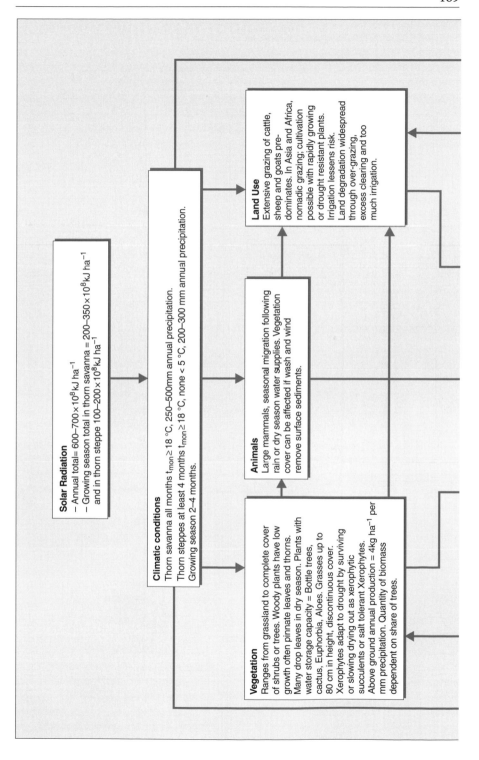

Solar Radiation
- Annual total= 600–700×10⁸ kJ ha⁻¹
- Growing season total in thorn savanna = 200–350×10⁸ kJ ha⁻¹ and in thorn steppe 100–200×10⁸ kJ ha⁻¹

Climatic conditions
Thorn savanna all months $t_{mon} \geq$ 18 °C, 250–500mm annual precipitation.
Thorn steppes at least 4 months $t_{mon} \geq$ 18 °C, none < 5 °C, 200–300 mm annual precipitation. Growing season 2–4 months.

Vegetation
Ranges from grassland to complete cover of shrubs or trees. Woody plants have low growth often pinnate leaves and thorns. Many drop leaves in dry season. Plants with water storage capacity = Bottle trees, cactus, Euphorbia, Aloes. Grasses up to 80 cm in height, discontinuous cover. Xerophytes adapt to drought by surviving or slowing drying out as xerophytic succulents or salt tolerant Xerophytes. Above ground annual production = 4kg ha⁻¹ per mm precipitation. Quantity of biomass dependent on share of trees.

Animals
Large mammals, seasonal migration following rain or dry season water supplies. Vegetation cover can be affected if wash and wind remove surface sediments.

Land Use
Extensive grazing of cattle, sheep and goats predominates. In Asia and Africa, nomadic grazing; cultivation possible with rapidly growing or drought resistant plants. Irrigation lessens risk. Land degradation widespread through over-grazing, excess clearing and too much irrigation.

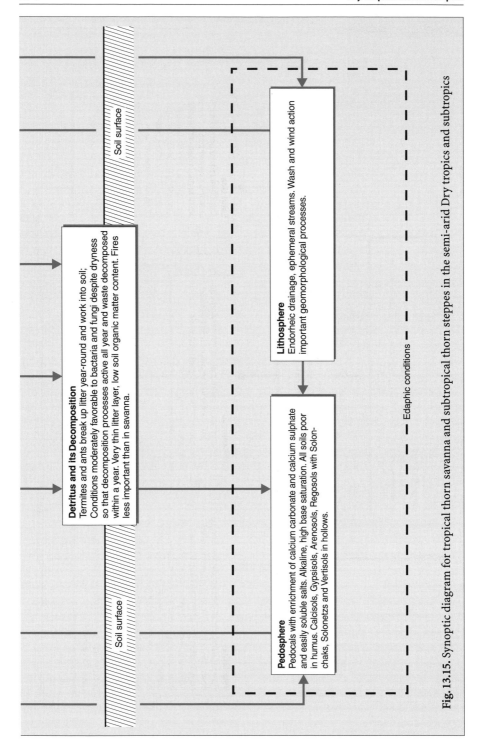

Detritus and its Decomposition
Termites and ants break up litter year-round and work into soil; Conditions moderately favorable to bactaria and fungi despite dryness so that decomposition processes active all year and waste decomposed within a year. Very thin litter layer, low soil organic matter content. Fires less important than in savanna.

Soil surface

Lithosphere
Endorheic drainage, ephemeral streams. Wash and wind action important geomorphological processes.

Pedosphere
Pedocals with enrichment of calcium carbonate and calcium sulphate and easily soluble salts. Alkaline, high base saturation. All soils poor in humus. Calcisols, Gypsisols, Arenosols, Regosols with Solonchaks, Solonetzs and Vertisols in hollows.

– – – Edaphic conditions – – –

Fig. 13.15. Synoptic diagram for tropical thorn savanna and subtropical thorn steppes in the semi-arid Dry tropics and subtropics

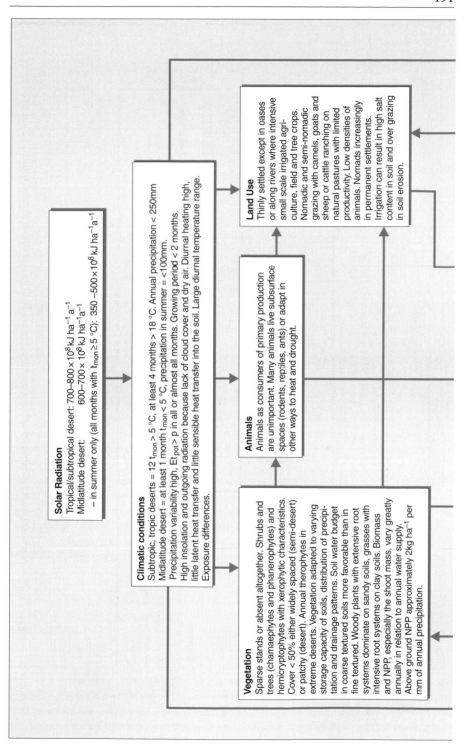

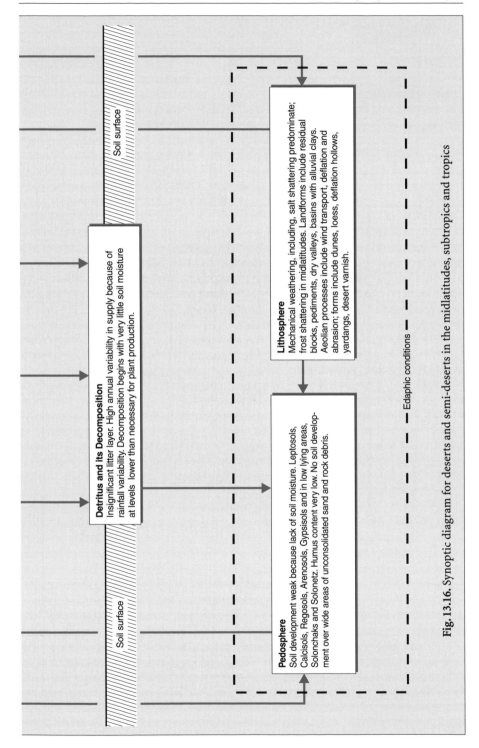

Soil surface

Detritus and its Decomposition
Insignificant litter layer. High annual variability in supply because of rainfall variability. Decomposition begins with very little soil moisture at levels lower than necessary for plant production.

Soil surface

Lithosphere
Mechanical weathering, including, salt shattering predominate; frost shattering in midlatitudes. Landforms include residual blocks, pediments, dry valleys, basins with alluvial clays. Aeolian processes include wind transport, deflation and abrasion; forms include dunes, loess, deflation hollows, yardangs, desert varnish.

Pedosphere
Soil development weak because lack of soil moisture. Leptosols, Calcisols, Regosols, Arenosols, Gypsisols and in low lying areas, Solonchaks and Solonetz. Humus content very low. No soil development over wide areas of unconsolidated sand and rock debris.

Edaphic conditions

Fig. 13.16. Synoptic diagram for deserts and semi-deserts in the midlatitudes, subtropics and tropics

CHAPTER 14

Tropics with summer rain

14.1
Distribution

The Tropics with summer rain lie between the tropical rainforests on the equator and the Dry tropics and subtropics. The boundary of the ecozone towards the arid regions is determined largely by the availability of water. An area is arid if the mean annual precipitation is less than 500 mm, of which nearly all falls in fewer than 5 months (Fig. 14.2). About 25 million km^2 or 16% of the land mass belong to this ecozone.

The main plant formations of the ecozone are various types of *savanna* vegetation. The terms savanna climate or savanna belt are sometimes used to describe the zone as a whole.

The savanna is subdivided into a *dry savanna* and *moist savanna* depending on the total annual precipitation and its duration (Figs. 1.1 and 14.2). The differentiation of vegetation, soils and land use within the savanna is also

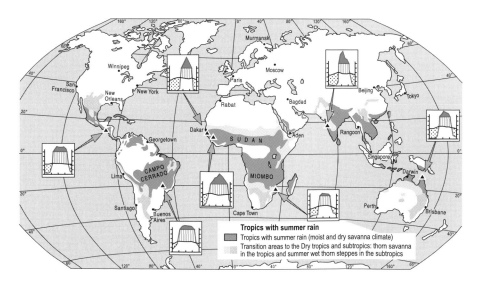

Fig. 14.1. Tropics with summer rain

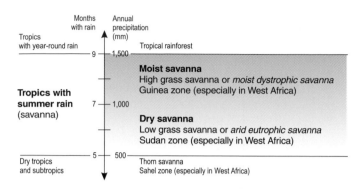

Fig. 14.2. Regional differentiation within the Tropics with summer rain

an expression of this climatic division. Grasslands have a considerably lower growth in the dry savanna, for example, compared to the high grass savanna of the moister areas where tree stands are also denser and higher than in the dry savanna.

Soils in the dry savanna usually have a higher exchange capacity and base saturation and are richer in humus than the moist savannas. The higher soil fertility allows a permanent or at least semi-permanent cultivation.

The soils of the moist savanna are usually developed on deeply weathered bedrock. Decomposition rates of organic detritus are high and leaching is considerable so that the soils are poor in nutrients and humus. Cultivation with periods of fallow and shifting cultivation characterize land use. The limited nutrient supply rather than lack of water determine the productivity of the soils.

The two savanna types are also known as arid eutropic savanna and moist dystrophic savanna, a reflection of the differences in soil fertility.

14.2
Climate

Because of the positive radiation balance throughout the year and small drop in temperature during the rainy season mean temperatures are equable year-round. The mean monthly temperature range is generally less than the diurnal range. In all months the mean is over 18 °C. The highest temperatures occur before the rainy season and the mean monthly maxima may exceed 40 °C at this time. Monthly means and also daily mean temperatures are lowest in the dry season; the minimum daily temperature occurs in the middle of the dry period (Fig. 14.3). Frost is absent during the rainy season but in some upland areas temperatures may fall below freezing in the dry season. The winter dry season lasts from 2.5 to 7.5 months. Annual precipitation ranges from 500–1,500 mm.

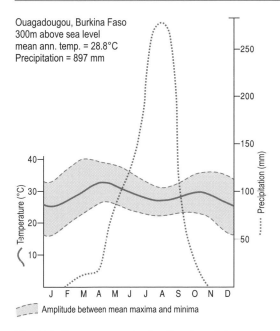

Amplitude between mean maxima and minima

Fig. 14.3. Temperatures in the Tropics with summer rain. An example of dry savanna at Ouagadougou, Burkino Faso. The dip in summer temperatures is caused by the cooling effect of the rainy season. Further north the dip is gradually reduced with the shortening of the rainy season and is not present in the Dry tropics and subtropics. The difference between the mean maxima and the mean minima is 9 °C in the rainy season in this example and 18 °C in the dry season because of the higher solar radiation during daytime and greater losses at night. Source: Müller 1996

14.3
Relief and drainage

Relief in the Tropics with summer rain is characterized by large areas of almost level land surfaces known as *peneplains*. Denudation has removed all differences in bedrock and geological structures and even valley sides have such a low angle that they are hardly visible in the landscape. The peneplains contain broad shallow drainage ways separated by broad, low overland flow divides. The shallow drainage ways, which are at most a few square kilometers in size, frequently serve as valley heads which fill with stagnant water or flood for a short period during the rainy season. In the dry savanna Vertisols are developed in these areas and in the moist savanna, Gleysols. Both are covered with grassland. In Zambia the hollows are known as dambos and in Tanzenia as mbugas.

On the divides and in the valleys *laterite* is often present at or a little below the surface as an irreversably hardened soil horizon of subsoil rich in iron (plinthite). This type of *duricrust* is termed *ferricrete* and often lies above the surface after wash denudation has removed the material at the surface

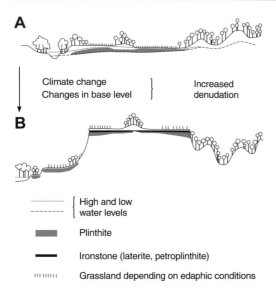

Fig. 14.4. The development of plinthite in areas of stagnant drainage or groundwater influenced surfaces and the irreversible hardening of plinthite to ironstone (laterite and petroplinthite). Plinthite develops when the ground is repeatedly dried out during a period of climate change from year-round rain to seasonal rain or when the base level changes A similiar effect can be created when forests are cleared and the soil dries out because of the increased solar radiation. Wash processes may cause further drying out of the plinthite horizon. Source: Spaargaren and Deckers 1998

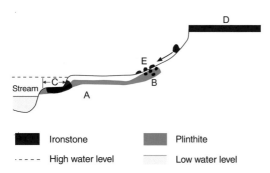

Fig. 14.5. Occurrence of plinthites and ironstone (petroplinthite) under moist tropical conditions in relation to relief. Soft plinthite develops in continuously moist subsurfaces of river terraces and flood plains (A) as well as areas near springs at the foot of uplands and scarps (B). If the plinthite dries out on, for example, river terraces (C) or wash surfaces of upland areas (D), irreversible hardening occurs and petroplinthite is developed in situ. Iron rich debris may accumulate in the colluvium on the foot slopes of uplands and scarps (E). Source: from Driessen and Dudal (1991)

(Fig. 14.4). Duricrusts also cement wash surfaces and form resistant layers in piedmont steps, on upland slopes or river terraces (Fig. 14.5).

During the rainy season much of the precipitation does not infiltrate but reaches the valleys as overland flow. Intense precipitation events are frequent in the tropics, sometimes in association with tropical cyclones and are the main cause, combined with the low *infiltration* capacity of the often, fine grained soils, that underlie the predominance of this process. Where the slope angle is low and the surface relatively smooth, sheet flow occurs, where the slopes are steeper and surfaces rougher, because of either boulders or vegetation, there is rill wash.

Over long periods of time all forms of denudation lead to a continual lowering of the *regolith* cover. The lowering of the underlying bedrock and the development of bas-relief surfaces, the actual peneplain development, occurs in conjunction with the chemical weathering process at the lower boundary of the regolith cover on the bedrock surface. On the land surface, the regolith is removed by wash denudation. The regolith cover advances the weathering process up to a certain depth because moisture in the regolith is stored in contact with the underlying bedrock so that the weathering continues between precipitation events and into the dry season. The lowering at the regolith surface and at the bedrock surface is termed double planation. When the rate of removal at the surface is the same as the rate of regolith production, a dynamic equilibrium exists. The often thick regolith cover in many areas of the tropics indicates, not only that weathering rates have been very high, but also must have been higher than the rates of denudation at the surface.

Rock surfaces in the form of *inselbergs*, *pediments* or piedmonts cover relatively small areas. They occur because the potential denudation has exceeded the ability of the rock to produce weathered material. Inselbergs are often isolated individual or groups of hills rising above a level surface. Their slopes are steep and the change in gradient to the plain often abrupt. Sometimes they are surrounded at the base by low angled pediments.

In some cases inselbergs are the result of tectonic activity, or when resistant rock was surrounded by areas of less resistant rock that was removed; sometimes they are isolated features a relatively large distance from streams or are exposed blocks that were more resistant to weathering at depth, perhaps because they were less jointed. Such blocks now rise above the surface as the weathered material that embedded them has been removed (Fig. 14.6)

Most smaller streams only flow in the rainy season. Stream flow at the beginning of the rainy season is associated with surface runoff from individual events and is therefore episodic with very high peak flows. With the advance of the rainy season flow becomes continuous, particularly as supplies from groundwater reach the stream. As the rainy season ends stream flow declines again to zero.

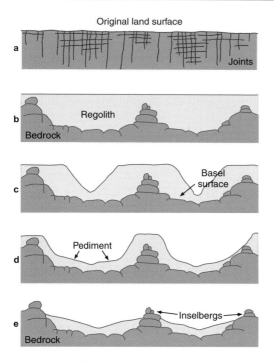

Fig. 14.6. Development of inselbergs. Source: Ollier 1984

14.4
Soils

14.4.1
General characteristics of soils in the subtropics and tropics with summer and year-round rain

The boundary area between the Dry tropics and the subtropics and the Tropics with summer rain is the equatorward limit of *pedocals*. When mean precipitation is above 500 to 600 mm, leaching replaces enrichment and *pedalfers* are the predominant soil type in well drained locations. They are acid and therefore poor in exchangeable nutrient ions. Free carbonates and salt are absent and the displacement of clay from the upper to the lower soil layers is widespread.

The pedalfers of the tropics differ from those in higher latitudes largely because of the warm moist climatic conditions and intensive chemical weathering. Characteristic for the tropics are two layer clay minerals, usually kaolinite, sometimes also halloysite. Sesquioxides are frequent, increasing with the intensity and duration of the warm wet conditions, reaching their maximum where silicium is released as a consequence of weathering. Hematites in particular develop following oxidation of the iron that is in many rock minerals, giving the soil a reddish color. There are few weatherable silicates in these soils. In extreme cases the delivery of mineral nutrients is almost entirely the result

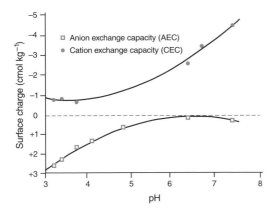

Fig. 14.7. The dependence of pH on the charging characteristics (anion and cation exchange capacity) in low activity clays in a ferralic B horizon (Ferrasol). Source: from Van Wambeke 1992

Table 14.1. Comparison of properties of selected soil types in the Subtropics and tropics with summer rain and the Subtropics and tropics with year-round rain

Soil type	CEC (cmol(+) kg⁻¹ clay)	Base saturation (%)	Illuviation of clay in B horizon	Characteristics
Lixisols	< 24	≥ 50	argic horizon	unstable soil structure
Acrisols	< 24	< 50	argic horizon	unstable soil structure
Alisols	≥ 24	< 50	argic horizon	high Al content
Nitisols	< 24	± 50	partial (nitic hor.)	favorable soil structure
Ferralsols	≤ 16	< 50	ferralic horizon (no enrichment)	micro sand aggregates
Plinthosols	< 16	< 50	plinthic horizon (no enrichment)	B horizon rich in iron, laterite development can occur

of decomposition of plant detritus and inputs from the atmosphere, both of which are concentrated in the uppermost soil layers.

All important zonal soils in the subtropics and tropics with high rainfall are low activity clay soils with a cation exchange capacity of less than $24\,cmol(+)\,kg^{-1}$ clay, they include Ferralsols, Plinthosols, Acrisols, and Lixisols. Nitisols are a border case (Table 14.1). The exchange capacities of low activity clays are more dependent on pH than high activity clays. In kaolinites, the surplus of negative charges increases with increasing pH, so the CEC increases also. On the other hand, the positive surplus charges in sesquioxides increase with declining pH, so that the AEC increases (Fig. 14.7).

One consequence of this is that in acid soils rich in sesquioxides, especially Acrisols and Ferralsols, the adsorption of phosphorous anions (PO_4), can be

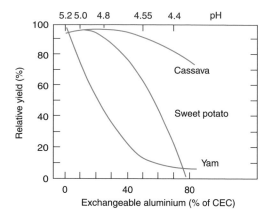

Fig. 14.8. Relationship between exchangeable aluminum, pH in the soil and yield of three tropical cultivated tubers. Source: Norman et al. 1995, Marschner 1990

irreversable and the phosphorous is fixed in the soil as iron phosphate and aluminum phosphate, creating very negative conditions for cultivation. Also, with increasing acidity, from about pH 5, the aluminum compounds tend to be dissolved and the aluminum solution in the soil can reach a saturation level that is toxic to plants, defined usually as a level above 60%. The tropical crop most affected by *aluminum toxicity* is the yam, sweet potatoes less so and cassava least of all (Fig. 14.8).

pH values can be improved by liming the soil which may also lead to an increase of up to 50% in the CEC. The disadvantages of phospate fixing and aluminum toxicity are reduced in this way or even disappear and the effectiveness of fertilization is greatly enhanced.

14.4.2
Soil types

Lixisols, Nitisols and Vertisols are the most common soils in the Tropics with summer rain. Other soils in the ecozone are discussed in the relevant chapters, Acrisols and Alisols in Chap. 12.4, and Ferralsols and Plinthosols in Chap. 15.4.

Lixisols formerly belonged to the soil group known as Luvisols. They are soils with a low CEC, in the B horizon sometimes less than $24\,\mathrm{cmol}(+)\,\mathrm{kg}^{-1}$. They also develop an argic B horizon and have a base saturation of at least 50%. The pH values and the quantity of exchangeable nutrient ions are higher than in Acrisols and Ferralsols. The yield potential is moderately high, despite low humus content and low activity clays. Lixisols are, however, vulnerable to soil erosion. After heavy rain the soil surface silts up and once dried out, crusts may form If heavy machinery or plows are introduced, the unstable soil structure is likely to be disturbed.

Nitisols are developed on silicate rich rocks such as basalt or schist and are relatively young. There are always remains of weathered materials in silt and sand grain sizes in the soils. The Nitisols are defined by a nitic horizon which has a stable structure and angular blocky aggregates with shiny surfaces. Although Nitisols have a high clay content (mainly kaolinite and sesquioxides) porosity is high enough for precipitation to infiltrate rapidly. A large amount of bound water can therefore be stored and the soil is well aerated. Nitisols are not greatly affected by soil erosion and have a high usable field capacity. Ecologically they are some of the best soils in the tropics and subtropics and traditionally have been cultivated permanently without fallow. Reddish to dark brown in color, they occur in isolated pockets and account for perhaps a fifth of the area within the more widespread Acrisols and Lixisols.

Vertisols are dark grey to black in color, except for the brown chromic Vertisols. Clay content is high, at least 30%, frequently more than 50%. Vertisols are one of the frequently occurring soils in the subtropics and tropics where there are from 3 to 9 dry months and at least 200 to 300 mm mean annual precipitation. They are developed on formerly grass covered level or gently inclined land surfaces and in poorly drained hollows on weathering products that have a high clay content containing primarily $CaCO_3$ and on sediments. In the rainy season, Vertisols have a sticky and plastic consistency but harden during the dry season that follows. Fissures and cracks form in the hardened surface that are from 1 to 10 cm in width and up to 150 cm in depth and which divide the soil mass into polyhedrons. The cracks become filled with soil material loosened and trampled in by animals or washed in during the first precipitation event after the dry season. In this way, the dry volume of the soil is increased and when subsequently wetted the soil swells causing pressure locally, at least in the subsurface. Movements within the soil follow during which the soil mass is homogenized and mixed. The development of an Ah horizon more than 1 meter in depth often results.

Movement is the defining characteristic of Vertisols. Within the soils, there are shear surfaces, slickensides, on which the clay minerals' surfaces lie parallel to one another and along which shearing takes place. At the surface, gilgai, a microrelief of small mounds and depressions, increase the possibility of movement in the soil because the distribution of preciptation on the surface leads to a small scale variation in the wetting of the soil. (Former names for Vertisols include regur and black cotton soils) (Fig. 14.9).

The high content of clay minerals in Vertisols, in particular smectite, underlies the swelling and shrinking and subsequent movement in the soil as the water content of the soil changes. The cation exchange capacity of 40 to 80 of the Vertisols is high and they are neutral to alkaline. Carbonates may be precipitated out, partly as concretions. Humus content is less than 3%, the humus material is in stable humus complexes and the carbon nitrogen ratio around 15.

The high nutrient content of Vertisols gives them a high production potential which, however, is limited by the problems created by the clay content. For example, the plants have a high wilting point because only a small proportion of the water stored in the soil can be used in the plants. Also losses from

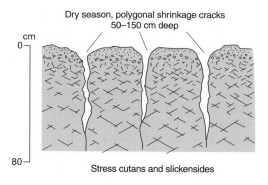

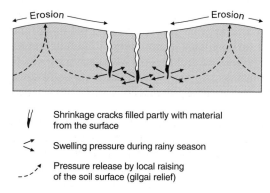

Fig. 14.9. Self-mulching and gilgai (micro) relief develoment in Vertisols. Source Driessen and Dudal 1991

evaporation are relatively high, in part because of capillary movement in the soil. Movement in the soil, in general, can affect root systems detrimentally. Moreover the soil is vulnerable to soil erosion and difficult to work because it is heavy and sticky when wet and very hard when dried out. Most Vertisols are used for pasture, either natural or semi-natural. Cotton, sugar cane, wheat and sorghum are cultivated but have to be irrigated.

14.5
Vegetation and animals

14.5.1
Structural characteristics of savanna vegetation

Savanna vegetation includes tree savannas, shrub savannas and grass savannas. The grass cover is uninterrupted but tree densities range from almost treeless areas to almost closed canopies. In most cases, the clearings are the result of fire, grazing or wood cutting and not related to the location. Studies in areas unaffected by human intervention indicate that the density, height and number of species increase in relation to the length of the period of non-intervention. It is likely that in the area of the moist savanna closed stands of trees originally covered far larger areas than at the present time. The grassland that dominates now is probably often a secondary vegetation.

The densities and heights of the trees are lower on soils with a high nutrient content in the dry regions of the savanna, the *arid eutropic savanna*, than in areas of greater moisture with soils having a low nutrient content, the *moist dystropic savannas* (Huntley and Walker 1982, Cole 1986).

Grass heights also vary. Grasses in the moist savanna reach more than 1 meter, sometimes even more than 2. In the dryer areas, the grasses are less than 80 cm high, but taller than the steppe grasslands in cooler climates.

The three to seven arid months are the most important factor in limiting plant growth. During the dry season, many tree species drop their leaves and the shoots of the perennial grasses and other herbaceous plants die off. The litter produced protects the growth point of the plants until the growth begins again in the following season In this way the photosynthesis decreases to zero or a little above for several months, similar to plants in the Temperate midlatitudes where activity is reduced in winter because of cold rather than the aridity.

When and for how long a tree in the savanna loses its leaves depends on the species and particular environment of the tree. Usually the *leaf drop* occurs several weeks after the herbaceous plants wither. If the water supply in individual years remains within the root area of the tree, the leaves do not fall in that year. The storage of water in most loamy soils is about 15–20% of the soil volume, 15 to 20 mm for each 10 cm in depth or 15 to 20 litres per square meter. The depth of storage depends on the amount of water surplus from the previous rainy season. The higher the surplus, the longer the growing period in any one year.

The leaves also change color before they fall, although with much less variation than in the deciduous trees of the midlatitudes. Many trees and shrubs bloom in the month before the rainy season begins.

14.5.2
Animals

Characteristic for the savanna is the wide range of insect life and spiders. Locusts, grasshoppers, roaches, hemipterans, beetles, flies, mosquitos, butterflies, bees, wasps, ants and *termites* are present in large numbers.

Termite hills range from less than one meter in height to 5 or 6 meters; they have a narrow or broad form and stand either singly or in dense groups covering up to 5% of an area. They are noticeable in the landscape not only because of their appearance but because vegetation near the termite hills differs from that in the surrounding areas. Tree and shrub cover is denser and the herbaceous and grass flora is not the same as elsewhere, such areas are sometimes described as termite savanna.

Among the large numbers of vertebrate species in the ecozone are the four main groups of reptiles, snakes, lizards, crocodiles and tortoises, all but the latter are carnivores. There are also large numbers of running birds, such as ostrich, nandus and emus and bustards as well as ground nesting birds and various birds of prey. Of the mammals, rodents, rabbits and hares are common in all areas.

The development of species on each continents has varied greatly. Not only do the groups of animals differ because of the environmental differentiation in the savanna in different parts of the world, but also because of the numbers of animals, the zoomass, and their growth form and function in the ecosystem. Many of the African savannas have a unique and greatly varied animal life.

14.5.3
Savanna fires

Few areas of the savanna have escaped *fires*. Caused for the most part by man, they usually occur in the dry season and accentuate the normal seasonality of the area. Fires can be advantageous or destructive in their effects. In an area immediately affected, fire is an important selection factor for fauna and flora. Fire also determines the vegetation structure, the tree cover for example, influences the heat and water budget of the soil, the plant cover and the air layer near the soil, as well as changes in the supply of materials and energy and turnover in the system, including the volume of above ground biomass and litter, turnover by herbivores and saprophytes, together with the recycling of minerals.

In general the annual emissions of nitric oxide (NO_2) ammonia (NH_3) and carbon dioxide (CO_2) from fires in the savannas are lower in quantity than the amounts bound into the biomass by photosynthesis and nitrification from the atmosphere during the year, or growing season immediately preceding the fire. Larger emissions can occur in the tree savannas when the woody plants have been burnt and have not regenerated.

Various stages of degradation and succession are recognizable in areas affected by fire depending on the frequency with which individual areas are burnt, whether the fire is at the beginning or later in the dry season and the length of the period since the last fire. The mosaic of vegetation in such areas, which is termed a *fire climax formation*, is a complex pattern parts of which may alter over time, but which, in the long term, is preserved basically. There will be changes in the structure only when fires no longer occur.

14.5.4
Biomass and primary production

The size of the *biomass* depends largely on the density and height of the tree cover. Since in most savannas the natural cover has been altered by human intervention, estimates of the total biomass do not indicate much about the ecology.

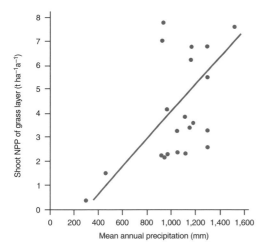

Fig. 14.10. Annual production of grass layer in the west African savanna in relation to the mean annual precipitation. The wide range of values reflect edaphic variations or variations in the density of woody growth at selected locations. Source: Ohiagu and Wood 1979

Biomass in the grassland savannas, as in the steppes, varies greatly with the season. During the dry season production can range from zero, immediately after a fire to the height of the net primary production towards the end of the rainy season. The actual amounts lie somewhere in between because not all the grass is burnt and part of the growth dies and drops off or is eaten by herbivores. The maximun herbaceous mass above ground is reached therefore before the end of growing season.

In addition to the seasonal variation, there are also considerable fluctuations from year to year because, particularly in the dry savanna, the production of grass in the years of high annual precipitation is much higher than in years with low rainfall totals. There is a fairly close linear relationship between the long term mean annual above ground primary production and the mean annual precipitation, similar to that in the Dry midlatitudes and the Dry tropics and subtropics. Figure 14.10 shows an example from west and central Africa. If South American savanna is included, a range of 5 to 7.5 kg of dry matter $ha^{-1} a^{-1}$ per millimeter of precipitation is estimated, a higher rain use efficiency therefore than the 4 estimated for semi-arid and arid regions (McNaughton et al. 1993; Chap. 13.5.4). If precipitation is higher, the linear relationship is not maintained because as the surplus of precipitation grows, more rainwater flows away unused, conditions that are similar to those in the moist savanna. The net primary production in the moist dystropic savanna is, therefore, at a similar level to the eutrophic savanna because the advantage of greater moisture is offset by lower soil fertility in moist areas. Where fertility is greater in the moist savanna, however, the net primary production and rain use efficiency increase with increasing precipitation (Werner 1991).

14.5.5
Zoomass and animal feed

Compared to other ecosystems in which herbivores eat less than 5 to 10% of the net primary production, in most savannas more than half the above ground herbaceous production and a quarter of the subsurface production is eaten by animals. Invertebrates consume more than the vertebrates. Grasshoppers and locusts are the primary consumers almost everywhere, together with caterpillars during the rainy season and roaches and crickets, which also eat dead matter.The most important secondary consumers include spiders and ants, many of which are omnivores. Among the detritus feeders, termites, ringworms, millipedes and the larvae of beetles predominate.

In some parts of the moist savanna in East Africa large mammals reach very high volumes of *zoomass* of from 0.1 to 0.3 t ha^{-1}, mostly elephants and hippopotamuses. Still 0.06 – 0.1 t ha^{-1} zoomass is found in some dry savannas of East Africa. In addition to grasses, browsing of trees and shrubs are the main source of feed for large mammals. Food shortages in the dry season may cause stress for herbivores when the herbaceous plants and leaves die off, or if fire destroys the plants. The availability of food remaining in the dry season and after fire plays a major role in determining which species are present in a region and their densities. Many animals migrate to offset the lack of food in the dry season by concentrating near water sources or in valley bottoms supplied by groundwater or by moving to areas where rainfall is more plentiful.

The biomass consumed by animals (C) is only partly assimilated (A) into their bodies. The rest is defecated, often in a relatively unaltered form and is available to detritus feeders (Chap. 5, box 4). The *assimilitation coefficient* (A/C) of herbivores varies greatly among species. It is also affected by the age of the animal and the type of food. The A/C of mammals lies between 30 to more than 60%, much lower than most carnivores but much higher than the detritus feeders (Table 14.2).

Of all the energy assimilated by mammal herbivores, from 1 to 5% is used for production (P), that is growth or reproduction. Cold blooded animals have a much higher net production efficiency (P/A) with generally more than 10% and occasionally over 50% of the assimilated energy being used for production.

The gross production efficiency or ecological efficiency (P/C) can be estimated from the assimilation and net production efficiency. Table 14.2 shows that the gross production efficiencies range from 0.5% for elephants to 50% for spiders, that is 0.5% of the intake of feed is used for production, including reproduction, compared to about half for spiders. Values lie between 1 and 10% for most animals with warm blooded animals having the lowest values. Compared to impalas and cattle, elephants require four times greater quantities of feed to produce the same growth.

Grazing of the grass cover by wild animals and cattle, the removal of the living biomass, does not affect primary production of the grass layer, rather it increases production capacity, particularly of the grass shoots, provided feeding is distributed over the growing season and plants are not weakened.

Table 14.2. Turnover efficiency and rates of turnover of selected animals in the savanna (%)

	Assimilation efficiency A/C	Net production efficiency P/A	Gross production efficiency P/C	Rate of turnover
Herbivores				
Locusts				
– Burkea savanna (various species)	32	19	6	
– Acacia savanna (various species)	32	21	7	
– Orthochtha brachycnemis	20	42	9	9.6
Caterpillar				
– Cirina forda	43	15	6	
Herbivrous and Saprovorous termites				
– Trinervitermes geminatus			9	10.4
– Ancistotermes cavithorax (fungi cultivating)			2	9.7
– Hodothermes mossambicus	61			
Ungulates				
– Waterbuck (Kobus kob)	84	1	1	0.27
– Impala (Aepyceros melampus)	59	4	2	
– Domestic cattle (Bos taurus) Transvaal.	57	5	2	
– African elephant (Loxodonta africana)	30	2	0.5	
Carnivores				
Spiders				
– Orinocosa celerierae	95	53	50	
Detrivores				
Earthworms				
– Millsonia anomala	9	4	0.6	2

A = assimilation, C = consumption, P = production

Source: from Lamotte and Bouliere 1983

Generally an intake as feed from the grass layer of from 30 to 45% of the net primary production can be tolerated.

The above ground NPP is raised because the plant matter consumed is altered to a more easily decomposable form which accelerates the return of minerals to the soil. In this way, plant growth is benefited rather than reduced. Also, shadow from lower leaf stories is decreased by browsing. If light is a growth

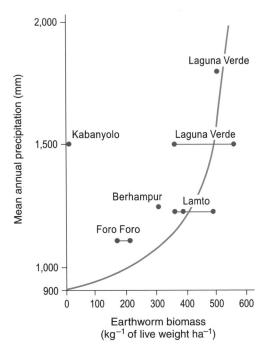

Fig. 14.11. Relationship between the size of the earthworm population and annual precipitation. Earthworms are active only when conditions are moist and are, therefore, most active in areas of high precipitation with long rainy seasons. During the dry season, earthworms move to the deeper soil layers. Source Lal 1987

factor, growth conditions are improved in this way too during the grazing season. Both compensate for the reduction of photosynthetically active parts caused by animal take off.

14.5.6
Decomposition of litter

Decomposition of organic matter is rapid if not by *fire* then, in the first instance, especially by *termites*. The highest densities of termites occur in Africa and Australia. Densities increase in relation to the amount of dead organic matter available from waste. The greatest production of waste is in the moist savanna where there are over 100 million termites per hectare with a live weight of $0.1\,t\,ha^{-1}$, compared to only a few million in the dry savanna. The proportion of waste consumed in relation to the total supply of waste remains more or less the same everywhere, unless fire intervenes to take over most of the decomposition process. From one fifth to one half of all waste is consumed by termites in these areas. Earthworms, leaf cutting ants in South America, millipedes and beetle larvae, the latter especially in Central America are the next largest consumers (Fig. 14.11). Fungi, actinomycetes and bacteria take care

Table 14.3. Nitrogen balance in two moist savannas

	Lamto (Ivory Coast) ($kg\,ha^{-1}\,a^{-1}$)	Venezuela (Trachypogon savanna) ($kg\,ha^{-1}\,a^{-1}$)
Input by rainwater	19 (4.5 anorganic)	2.6
Biological N_2 fixation		
– Blue green algae	–	0.7
– Microorganisms in root area	9–12	6.7
Loss by fire	17–23	8.5
Leaching	5.6	0.5
Balance	+3.9	+1.0

Source: Medina 1987

of the final decomposition stage, returning the organically bound minerals to a single anorganic form which is then reavailable to the plants. This process of recycling, in effect the duration of the decomposition, usually takes place within a year, despite slowing during the dry season.

Humus formation in the savanna is relatively unimportant and the humus content in the soils, therefore, relatively small, as are supplies of nitrogen and phosphorus, which can be a limiting factor in production.

14.5.7
Mineral supplies and turnovers

The supply of *plant nutrients* in the soil is generally unfavorable, particularly in the moist savanna. It is important that the existing nutrient supply is maintained and losses are replaced. Grasses for example have a high ratio of roots to shoots, dense root systems connected with mycorrhiza and nitrogen fixing Azospirillum. Rhizobium, another nitrogen fixing bacteria occurs in root modules of most leguminous trees and herbs.

Nitrogen fixing and the input of nitrogen by rainwater equal, in the long term, the losses resulting from leaching in the soil and from the burning of organic matter during fires (Fig. 14.3). Phosphate losses are generally replaced from the atmosphere.

Forests have a lower mineral requirement per unit of growth (i.e. higher nutrient efficiency) than grasslands. On the other hand, the circulation of minerals in the trees is much slower and therefore much larger quantities of minerals are bound in their biomass. Thus, their apparent advantage resulting from their higher NUE can be ignored. The requirements regarding soil fertility for grasslands in similar locations may be higher, but the proportion of the total mineral supplies which is available for uptake is larger in grasslands.

14.6
Land use

The Tropics with summer rain are the most densely settled and agriculturally most intensively used areas of the tropics, with the exception of some former rainforest areas of Southeast Asia. Compared to the Tropics with year-round rains, the ecozone has several advantages. Soil fertility is higher and there is a winter dry period which allows clearing by fire in areas still covered by woodlands. Extensive grasslands are available for cattle grazing also, the high intensity of solar radiation at the end of the rainy season is advantageous for the cultivation of maize, sugar cane and cotton.

In most summers in the Tropics with summer rain the total precipitation and the duration of the rainy season are sufficient for the cultivation of a wide variety of crops: maize, sorghum, millet, cotton, peanuts, rice, various varieties of beans and sweet potatoes. Because of a dry period of at least three months, only annual plants are grown unless they can be irrigated, as in the case of sugar cane, or are relatively more drought resistant such as cassava or sisal. Coffee and tea are planted in upland areas where orographic rainfall and low cloud provide moisture during the dry season.

The cultivation of a range of crops combined with some animal husbandry is typical for the predominantly small farms in the ecozone. The integration of crop growing and the keeping of animals has traditionally not been strong but has increased as more animals have been used to pull plows and their dung used to improve soil. Feed crops are absent although the cattle, sheep and goats feed in the fields after harvest and on the fallow areas, or on any available natural pastures held in common.

Semi-permanent cultuvation with land rotation is still the traditional form of agriculture in many areas. Fields are used for several years and are then left as *fallow* for the soil to regenerate. Some use is made of fallow in grazing areas.

A measure of the relationship between the number of years a piece of land stays in cultivation and the number of subsequent fallow years can be expressed as a cultivation factor. For example if a five year cultivation period is followed by five years of fallow, the cultivation factor is 2. In this case a farmer requires two areas of similar size and quality to maintain the level of output on his farm.

A rotation factor is another measure in which the cultivation years are expressed as a reciprocal of the total area used for cultivation, usually $\times 100$. Five years of cultivation and five years of fallow would result in a R factor of $5/10 = 0.5 \times 100 = 50\%$, indicating an area for crop use of 50% and a fallow area of 50%. A land rotation system has usually a factor of $0.3 \leq R \leq 0.7$. In this case a subsistence farm would need $1\,^1/_2$ to 3 times more land than a similar farm with permanent arable cultivation. Where soil conditions are poor and a long fallow is needed to regenerate the mineral nutrient budget, even more land is required. This is true for many areas in the moist savanna and the tropical rainforests (Chap. 15.6). Figure 14.12 shows soil productivity in relation to the length of the fallow period in shifting cultivation over a period of 40 years.

Production on the traditional mixed farms is mainly subsistence. Hoe and plow drawn by oxen are still widespread. Apart form some irrigated areas productivity is also low.

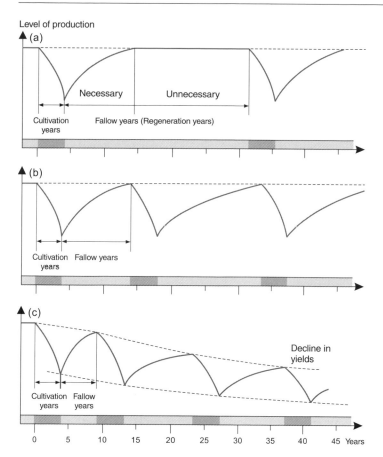

Fig. 14.12. Relationship between length of fallow and soil productivity in areas of shifting cultivation. In (b) the fallow period is equal to the period required to regenerate the productivity. In (a) the fallow is longer than necessary, in (c) the fallow period is not long enough to restore productivity and the yields per hectare decline. The figure also shows which rotation and fallow sequence brings the highest production over a longer period. With a shortening of the fallow period there are more harvests, which in time might compensate for the drop in yields per harvest, and this could in the long run – as long as the reduction of fallow periods is moderate – result in an increase of total production. Source: Ruthenberg 1980

Population growth and the consequent shortage of land has led to the replacement of the extensive shifting cultivation by a more permanent field system. An increased use of fertilizer has made this possible. Improved agricultural use of land within the forests has also reduced soil erosion and raised the quality of the organic content of the soil.

A major exception to the traditional extensive land use is the irrigated rice cultivation of Southeast Asia which covers large areas in both the Tropics with summer rain and the Tropics with year-round rain (Fig. 6.2).

Within areas of traditional cultivation, including irrigated areas, there are also commercial farms that specialize in the production of one or a few crops on relatively large arable farms. Maize, sorghum, tobacco, peanuts, cotton, wheat, coffee, tea and sisal are among the most frequent crops grown for commercial production. Cattle fattening and dairy farms have also developed. Commercial farms are often located along highways or form islands within an area of subsistence agriculture

In some areas of the savanna, in northern Australia, South America (Columbia, Venezuela, Paraguay and Brazil), Mexico and Kenya and Angola commercial cattle grazing has become established on large areas of unsettled land. Commercial grazing of wild animals has been attempted in several areas, including kangaroo in Australia, eland antelopes in Africa and capybaras in South America.

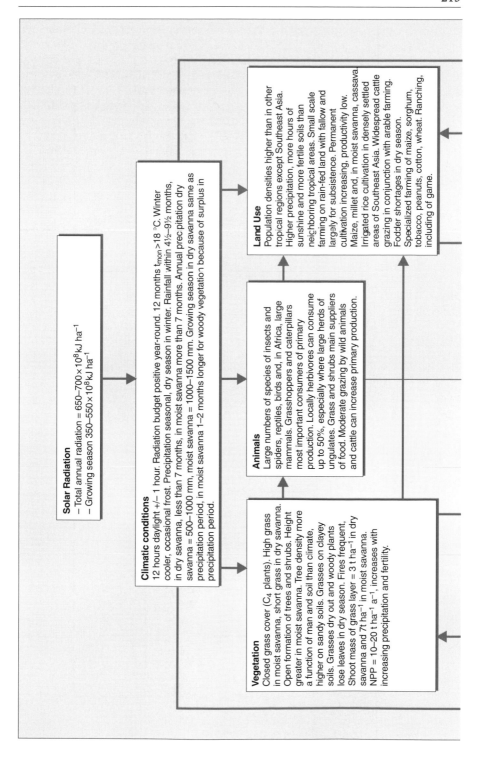

Solar Radiation
- Total annual radiation = 650–700 × 10⁸ kJ ha⁻¹
- Growing season 350–550 × 10⁸ kJ ha⁻¹

Climatic conditions
12 hours daylight +/– 1 hour. Radiation budget positive year-round. 12 months t_mon >18 °C. Winter cooler, occasional frost. Precipitation seasonal, dry season in winter. Rainfall within 4½–9½ months, in dry savanna, less than 7 months, in moist savanna more than 7 months. Annual precipitation dry savanna = 500–1000 mm, moist savanna = 1000–1500 mm. Growing season in dry savanna same as precipitation period, in moist savanna 1–2 months longer for woody vegetation because of surplus in precipitation period.

Vegetation
Closed grass cover (C₄ plants). High grass in moist savanna, short grass in dry savanna. Open formation of trees and shrubs. Height greater in moist savanna. Tree density more a function of man and soil than climate, higher on sandy soils. Grasses on clayey soils. Grasses dry out and woody plants lose leaves in dry season. Fires frequent, Shoot mass of grass layer = 3 t ha⁻¹ in dry savanna and 7 t ha⁻¹ in moist savanna. NPP = 10–20 t ha⁻¹ a⁻¹, increases with increasing precipitation and fertility.

Animals
Large numbers of species of insects and spiders, reptiles, birds and, in Africa, large mammals. Grasshoppers and caterpillars most important consumers of primary production. Locally herbivores can consume up to 50%, especially where large herds of ungulates. Grass and shrubs main suppliers of food. Moderate grazing by wild animals and cattle can increase primary production.

Land Use
Population densities higher than in other tropical regions except Southeast Asia. Higher precipitation, more hours of sunshine and more fertile soils than neighboring tropical areas. Small scale farming on rain-fed land with fallow and largely for subsistence. Permanent cultivation increasing, productivity low. Maize, millet and, in moist savanna, cassava. Irrigated rice cultivation in densely settled areas of Southeast Asia. Widespread cattle grazing in conjunction with arable farming. Fodder shortages in dry season. Specialized farming of maize, sorghum, tobacco, peanuts, cotton, wheat. Ranching, including of game.

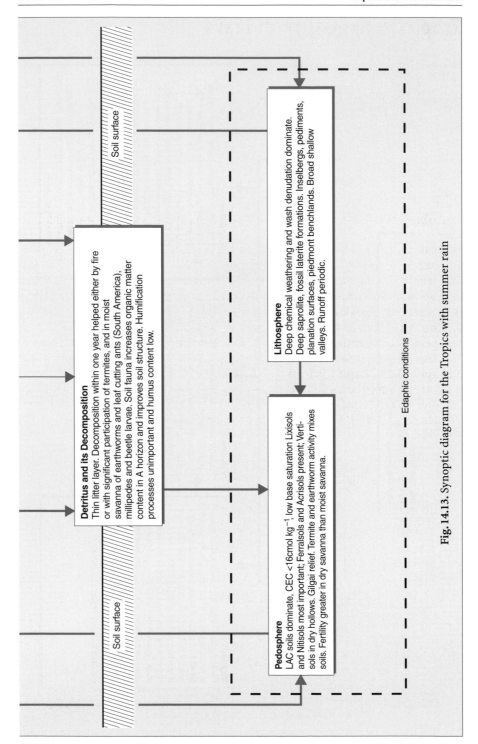

Detritus and its Decomposition
Thin litter layer. Decomposition within one year helped either by fire or with significant participation of termites, and in moist savanna of earthworms and leaf cutting ants (South America), millipedes and beetle larvae. Soil fauna increases organic matter content in A horizon and improves soil structure. Humification processes unimportant and humus content low.

Lithosphere
Deep chemical weathering and wash denudation dominate. Deep saprolite, fossil laterite formations. Inselbergs, pediments, planation surfaces, piedmont benchlands. Broad shallow valleys. Runoff periodic.

Pedosphere
LAC soils dominate, CEC <16cmol kg⁻¹, low base saturation Lixisols and Nitisols most important; Ferralsols and Acrisols present; Vertisols in dry hollows. Gilgai relief. Termite and earthworm activity mixes soils. Fertility greater in dry savanna than moist savanna.

Soil surface

Edaphic conditions

Fig. 14.13. Synoptic diagram for the Tropics with summer rain

CHAPTER 15

Tropics with year-round rain

15.1
Distribution

Most of theTropics with year-round rain lies within latitudes 10°N and 10°S. It extends, however, both north and south of the equator to about latitude 20° where the ecozone is in the area in which the winter trade winds and monsoonal rains, mostly orographic rains, supplement the summer rainfall. The total area is about 12.5 million km², 8.4% of the land mass.

The boundary to the Subtropics with year-round rain is determined by temperature and runs along the 18 °C isotherm for the coldest month.Towards the Tropics with summer rain, the boundary is determined by a mean annual precipitation of about 1,500 mm. Soil, relief, vegetation and land use have some similarities with the moist savanna zone of the Tropics with summer rain. This justifies to a certain degree a combination of both which then is termed *humid tropics*. On Fig. 15.1 the moist savanna zone is shown as a transition zone to the Tropics with year-round rain.

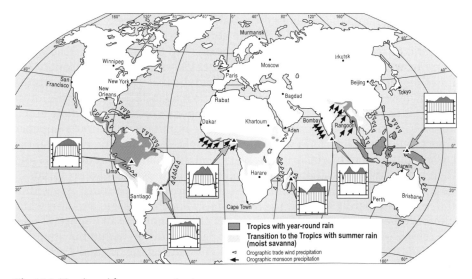

Fig. 15.1. Tropics with year-round rain.

15.2
Climate

Rainfall is year-round, especially near the equator where there is no noticeable seasonality at all. The strongly positive radiation balance is almost the same throughout the year because of the uniform length of day. Temperatures remain constant year-round at 25 to 27 °C, with a diurnal range of 6 °C to 11 °C. Convection rainfall predominates in the zone. There are two seasonal maxima shortly after the sun is at its highest point in April and October. In the months between precipitaion declines but there are seldom more than 2 to 3 months without any rainfall. Plant growth continues throughout the year, although in many areas it slows for a short period during which leaves fall or flowers develop.

Individual precipitation events are intense, short thunder showers are frequent and daily totals often more than 50 mm. Mean annual precipitation ranges from 2,000 to 4,000 mm. Annual cloud cover is often over 60% and 40% of the global radiation is diffuse radiation, the highest ratio of all ecozones. Much of the radiation intake is used for evaporation and transferred as latent heat. Relative humidity is very high throughout the year. Annually, evapotranspiration consumes over 1,000 mm, maximally over 1,200 mm, of water, more than any other ecozone and more than from the open sea in these latitudes. The plant surface area of the rainforest is very large and the supply of moisture from the soil is uninterrupted, even where there is a regular dry season because moisture continues to rise from the deep roots. There is a continuous and large amount of energy available for evaporation throughout the year and total actual evaporation almost reaches the potential evaporation.

In the areas of tropical forest, the climatic characteristics valid for the ecozone as a whole apply mainly to the canopy and the area immediately above it. The climate of the trunk area within the forest and particularly the climate in the air layer immediatly above the ground vary considerably for a number of reasons. First, sunlight is diminished to about 1 to 3% on the forest floor also, the ratio of red/infrared radiation energy sinks to 0.4 and less, instead of 1.0 or more, which affects germination and other growth processes. Secondly, the mean air temperature is equable with a small diurnal range of 3–4 °C. In the canopy, however, the midday temperatures reach up to 32 °C but near the forest floor it is usually 4–7 °C less. At night the temperatures are everywhere about 20–22 °C. Thirdly, the constantly high relative humidity of 90–100% impedes transpiration and interferes with mineral uptake from the soil, except on the surface of the canopy which at times has a relative humidity of only 50% and therefore a high saturation deficit. Leaf temperatures can rise up to 40 °C. Fourthly, the winds that accompany thunderstorms are felt much less within the forest and hardly at all on the forest floor. For this reason, and also while photosynthesis processes and the uptake of CO_2 do not play an important role, the release of carbon dioxide as a result of decomposition within the litter and soil leads to a concentration of CO_2 of 400 ppm during the day and 450 ppm at night; in the canopy, values during the day sink to nearly 300 pp (Fig. 15.2).

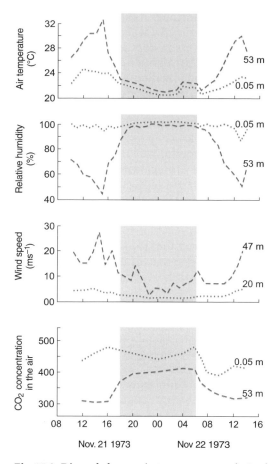

Fig. 15.2. Diurnal changes in temperature, relative humidity, wind speed and CO_2 concentration in the atmosphere above the canopy and in the trunk area in a lowland rain forest near Pasoh in western Malaysia. The surface of the closed canopy is at 45 m. Emergent trees reach 55 m. The leaf area index is 8. Source: Aoki et al. 1975

15.3
Relief and drainage

Chemical weathering is of major importance in the tropical rainforests. High soil moisture content, soil acidity and year-round high temperatures accelerate the process. Also, temperatures in the soil can be a few degrees higher than the air temperature near the ground because of heat released during humus formation.

Mechanical weathering is unimportant. Any disintegration of bedrock is a result of pressure release. Unlike the dry and and seasonal tropics, there is little material prepared by mechanical weathering for chemical weathering. Bare rock, especially if there are no joints and the exposures are steep is not decomposed.

Very deep layers of soil and *regolith* have formed over very long periods, often on old land surfaces, as a result of chemical weathering. Below the regolith is saprolite, decayed rock up to 100 meters in depth that lies above the bedrock. The regolith contains hardly any of the minerals present in the original bedrock. Relatively stable secondary products dominate such kaolinite, gibbsite, hematite and goethite so that there is little solution transport in the streams.

Solution processes in areas of limestone in the tropics have led to the formation of karst landscapes. Tower karst and cone karst up to 100 meters in height are characteristic for areas of the tropics with rainfall during much of the year.

Denudation surfaces are widespread in the Tropics with year-round rain, especially in areas of residual relief. Where there are young fold mountains and volcanoes, slopes are steep and incised by streams in valleys that are separated by narrow ridges. Because of the high precipitation, stream network densities are also high.

Little or no wash denudation can take place in the tropical forests. The rain falls on the stories of the canopy of the rainforest and, after a period of time, from 5 to 50% reaches the ground as drops, much of it as through flow and enters the soil. Moisture not taken up by the roots flows to the groundwater.

Mass movements are relatively frequent in the ecozone. Landslides, mud and earth flows all occur if the regolith becomes saturated and positive pore water presssure develops within the mass on the slope. Even though this form of denudation process takes place irregularly, it is the most important form of slope denudation in the tropical rainforests. Over time all slopes are probably denuded by mass movements which then reoccur in the same place when the regolith cover has developed again.

15.4
Soils

Some soils in the Tropics with year-round rain are similar to the soils in the Tropics with summer rain and the Subtropics with year-round rain. Ferralsols are the most widespread soil type. Plinthosols, Ferralic Cambisols, Ferralic Acrisols and Podzols are also present. Acrisols are common in Southeast Asia, West Africa and in the tropical rainforest of South America (Chap. 12). Lixisols are also important in some areas (Chap. 14). Table 15.1 shows the characteristics of the soils most typical for the ecozone.

Ferrasols are found for the most part on older land surfaces. They have developed in tropical rainforests over a very long time, often beginning in the Tertiary, on various bedrocks and under continuously wet and warm conditions. Acrisols have a similar but less advanced development.

Ferralsols are from light yellow to deep red and have a low humus content. The ferralic B horizon is very deeply developed, uniform in color and texture and without clay accumulation. Clay immobility is typical of Ferrasols. Weatherable silicate is less than 10% in the 50 to 200 µm fraction of the B horizon and the texture is sandy loamy and fine grained with clay forming about

Table 15.1. Chemical characteristics of soils in the humid tropics at the surface (upper 0–20 cm) and in the subsoil (70–100 cm)

Chemical characteristics	Ferralsols	Acrisols	Lixisols	Cambisols	Arenosols	Podzols
pH H$_2$O (1:2.5)	4.8 (5.0)	4.8 (4.8)	6.4 (5.9)	5.3 (5.5)	5.3 (5.8)	4.5 (4.8)
pH Kcl (1:2.5)	4.1 (4.5)	4.1 (4.0)	5.5 (4.6)	4.6 (4.5)	4.1 (5.1)	3.7 (4.4)
Organic carbon (%)	2.3 (0.4)	2.0 (0.4)	62.2 (0.3)	2.3 (0.4)	0.8 (0.1)	5.0 (0.7)
C/N ratio	16 (9)	14 (8)	17 (7)	11 (8)	16 (12)	23 (11)
Exchangeable bases (cmol(+)kg^{-1})	1.8 (0.7)	2.2 (0.6)	21.2 (16.8)	11.5 (9.0)	2.0 (2.0)	1.0 (0.1)
Exchangeable Al (cmol(+)kg^{-1})	1.4 (1.1)	1.5 (2.2)	0.0 (0.3)	0.1 (0.0)	0.1 (0.0)	1.0 (0.2)
CEC$_{pot}$ (cmol(+)kg^{-1})	8.8 (4.0)	9.9 (6.9)	22.7 (25.0)	19.3 (14.9)	6.6 (3.2)	20 (4.7)
Base saturation (%)	19 (19)	26 (12)	87 (67)	49 (52)	44 (39)	18 (43)

Values in parenthesis for subsoil
The values are means of 30 Ferralsols, 33 Acrisols, 9 Lixisols, 30 Cambisols, 5 Arenosols and 6 Podzols.
Exchangeable bases: Ca^{++}. K$^+$, Mg^{++} and Na$^+$

Source: Kauffman et al. 1998

8% of the fine earth fraction. The clay fraction is mostly of kaolinite, iron and aluminium oxide and hydroxide and silt/clay ratios are low. Most of the sand and silt particles are of quartz. Ferrasols are acid to very acid with a low cation exchange capacity and base saturation. (CEC$_{pot}$ < 16 cmol(+) kg^{-1} clay and CEC$_{eff}$ < 12 cmol(+) kg^{-1} clay). Ferralitization is the soil process that leads to the development of a ferralic B horizon. Included is the desilification process in which silicate is mobilized and washed out and without which the residual enrichment of kaolinite and sesquioxides cannot take place.

Iron (Fe) and aluminum (Al) oxides have a tendency to develop stable soil aggregates of sand sized grains (pseudosand) from a few millimeters to centimeters in size. Ferrasols with a high clay content have, therefore, a large proportion of stable intergranular large pores which makes them highly porous. Even after heavy rainfall they are not easily eroded and can still be worked. Of the water remaining in the soil, a large part is dead water within the soil aggregrates and not usable for plants. The maximum amount of usable water is usually less than 100 mm per one meter in depth (Spaargaren and Deckers 1998). Despite the year-round humid climatic conditions, shallow rooted plants are subject to stress from drought if there is a lack of rainfall for some time.

Plinthic Ferralsols have an upper B horizon of *plinthite*, a mixture of kaolinite, sesquioxides and quartz, high in iron and with a low humus content. The concentration of iron is a residual enrichment and appears in red flecks, The kaolinite concentrates as white flecks. Soils are firm in consistency and the porosity poor. Denudation rates on the slopes of these soils are high but on level ground they cause stagnation and flooding. If there is repeated drying out on fields after the top soil has been removed, the plinthite hardens as crust or as individual aggregrates, known as *ironstone*, that is irreversible. Formerly these soils were described as *laterites*.

The new FAO classification defines soils with a horizon at least 15 cm thick within 50 cm of the soil surface that has more than 25% plinthite and therefore a plinthic horizon, as Plinthosols.

Ferralic Cambisols are an indication of an early stage of ferralitic weathering. Unlike the Ferrasols, they contain unweathered minerals and have a higher cation exchange capacity than Ferrasols but lower than other Cambisols. Ferralic Cambisols are relatively fertile and in West Africa are developed in hilly regions where the top soil has been repeatedly removed by wash denudation so that soils remain in the early stages of formation. Leptisols are also often in these areas.

Ferralic Arenosols are sandy soils with a kaolinite clay component. In the tropics where rainfall is high, they are probably developed on relatively recent deposits of material that originated in highly weathered soils such as Ferrasols and Acrisols. The silicate content is low also the cation exchange capacity and the water storage capacity. The field capacity is often only about 10%.

Podzols are developed on sandstone and sediments with a high quartz sand content, often well over 80%, similar to the Arenosols. They are some of the poorest soils in the tropics and unlike all other soils in the zone, contain a high proportion of low quality dead organic matter with a low C/N ratio. The quantity of exchangeable nutrient ions is smaller than in many Ferralsols. Podzols are developed in the area of the Rio Negro in Brazil and on the Indonesian islands of Kalimantan and Sumatra, but cover in all only about 1% of the tropics with high rainfall.

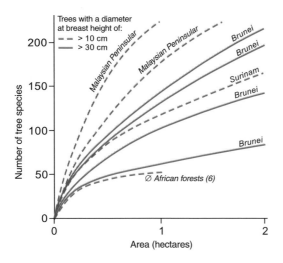

Fig. 15.3. Number of tree species per hectare. The curve for Africa is based on the mean of 6 surveys. The other curves are based on one survey

15.5
Vegetation and animals

The *evergreen tropical lowland forest* is the typical plant formation in the ecozone.

Clearing, especially during the last twenty years has reduced the area by more than half. With clearing, large amounts of *carbon dioxide* (CO_2) are released from the biomass and the humus. Even when CO_2 is bound again in crops, plants, pastures, forestation or secondary forest and the supplies of organic matter in the soils recover after a few years, there is still an overall loss. Whether this in reality reaches a total of $1.9\,GtCa^{-1}$ as shown in the global balance of the global carbon circulation is uncertain (Fig. 8.13). The exact total forest cover destroyed is unknown, also it is very difficult to estimate how much CO_2 is bound when the forest regenerates or clearings are cultivated.

15.5.1
Structure of the tropical rainforest

Tropical rainforests have a large number of species, more than one third of all known species in the world are from the tropical rainforests. There is also a wide *diversity of species*. For trees alone the number species per hectare can reach over 100, with at most only 2 or 3 trees of a single species in the hectare (Fig. 15.3). A comparison of the height and density of the plant coverage in the tropical rainforest compared to the Temperate midlatitudes is shown in Fig. 15.4. In the tropics there are several hundred trees, maximally about 1,000, with a trunk diameter of at least 10 cm at 1 meter above the ground. The total basal area of all trunks is at least 25, usually 30 to 40 $m^2\,ha^{-1}$.

The *leaf area index* in tropical rainforests at 8 to 12 is very high. The leaf stories in the forest can be measured for leaf density using the leaf surface area in m^2 per cubic meter volume of stand or the biomass (leaf and/or wood mass) per meter of height in $kg\,m^{-3}$ (Fig. 15.5). Each story has light intensity characteristics and other climatic parameters that vary from the means for the forest as a whole.

Often over 70% of the species in a forest are broad-leaved trees of which nearly all are evergreen. Both trees and herbaceous plants are hygrophytes and adapted therefore to the constantly high humidity. In addition to the trees, there are many species of lianas and vascular epiphytes. *Lianas* are woody climbing plants that reach considerable heights and use relatively limited nutrient supplies because they are supported by other plants. Their share of total basal areas of the trunks is low but at the canopy level their share of the leaf mass is high (Table 12.1). Over 90% of the liana species are in the tropical rainforests where they also provide a large share of the total flora, both in variety and quantity.

Epiphytes are usually herbaceous plants which grow on the trunks or branches of trees but without any indication that they are parasites. They include many orchids, the majority of which are epiphytes. Also widespread are ferns and

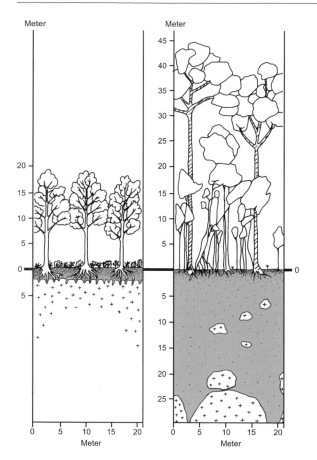

Fig. 15.4. Profile of a deciduous forest in the Temperate midlatitudes and of a tropical rain forest. Characteristic for the rain forest is the height of the trees, dense stands with large diameters, several levels of tree growth, underdeveloped herb layer, incomplete litter layer and, in relation to tree height, shallow rooting systems in humus poor deep soil that lies above decomposed bedrock

Bromeliaceae, mosses and lichens. Figure 15.6 shows the nutrient sources of epiphytes. The tallest trees in the canopy and many of the epiphytes are mesophytes or xerophytes.

Many broad *leaves* in the tropical rainforest are undivided in the middle, unlike leaves in the savanna. They are also larger, from 10 to 20 cm long, softer and darker green than leaves in other ecozones. The large leaves of many species encourages photosynthesis in limited light conditions. Some leaves also have hydathodes through which water is actively pressed into the saturated air. In many varieties the stomata are raised above the leaf surface and the total pore area, the product of the pore density per mm^2 and maximum pore width, is up to 3% of the leaf surface, higher than anywhere else. Frequently the leaves

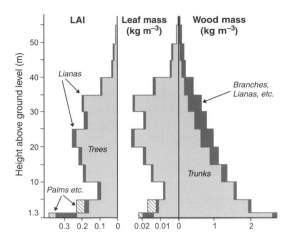

Fig. 15.5. Vertical structure of rain forest near Pasoh, Malaysia. Values at 5 m intervals of leaf area index, the leaf mass (kg per cubic meter), and, less defined, the wood mass of the branches and lianes, together indicate three forest stories at < 1.3 m, 20–25 m and 30–35 m. The rapid decrease in values above 40 m coincides with the story above the canopy composed of emergent trees. Source: Kato et al. 1978

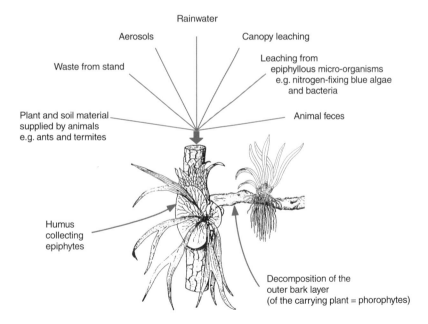

Fig. 15.6. Origin of the root substratum and nutrients for epiphytes. Source: Johansson 1974

have long pointed drip tips. In Borneo, Sri Lanka and Nigeria, 90% of the tree species have this type of leaf which it is assumed accelerates the runoff of water after rainfall so that the exchanges of gas necessary in photosynthesis and respiration are facilitated.

In the canopy conditions are quite different. Many hours daily of direct solar radiation and exposure to wind can subject the canopy to temporary drought stress. Leaves are xeromorphic, smaller than on species within the forest and leathery. Transpiration is slowed by a wax layer on the leaves and thick cuticles

The growing points for leaves and blossoms are not protected and bud scales are absent, at least on leaf buds. Protection has not been developed against cold or heat as in other ecozones. Where there is protection it is to prevent animals feeding on the buds.

In a large number of species of trees, the shoots and leaves grow rapidly, up to 20 to 30 cm a day. This is possible because hardly any structural tissue and chlorophyll are formed. Hanging shoots are, therefore, white or reddish in color initially. Many plants are cauliflorous. The flowers and fruits of these species grow on the main trunks from leafless shoots, for the cocoa tree and jackfruit this growth form is advantageous because of their heavy fruits.

Buttress roots in which the trunk and roots near the soil are enlarged are also characteristic of the rainforest, in some areas up to 40% of the trees have this root form which supports respiration.

Unlike all other ecozones there is no clear periodicity and no conspicuous visible change apparent in the tropical forest during the year, except where there is a short dry period of up to three months. Seasonally related leaf development, flowering, fruiting and leaf drop and also ring development in the trunks are all absent. These processes are, however, not continuous but occur as phases between long rest periods. The development phases occur at different times of the year and rarely coincide with seasonal changes in the climate. The flowering periods of different species are spread throughout the year. Even within the same species or on different branches of a tree, flowering is not simultaneous.

15.5.2
Dynamics of the vegetation

Each rainforest is a mosaic of stands of different ages which have a distinct fauna, flora, structure, available supplies and turnovers (Fig. 15.7). None is in a steady state. New development takes place when old trees have died off and fall so that gaps occur in the forest in which new growth can be established.

15.5.3
Animals

In most tropical rainforests animals are rarely seen and also hardly heard.The total number of species is larger than in any other ecozone although many species are represented only in small numbers and distribute themselves in a wide variety of ecological niches. The number of reptiles and amphibiens which thrive in the optimal conditions of temperature and humidity is particularly high as is the number of invertebrate species. Most animal species live

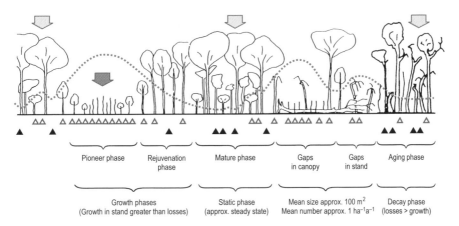

Fig. 15.7. Transect through a rain forest showing stages of maturity of the stand. There are four stages from pioneer to aging. During these phases the mix of species, the relationship of the NPP to waste in the stand and the availability of nutrients in the soil all change. The mature phase in undisturbed forest accounts for about 90% of the total area of the stand. The immature area covers 3–10% and the gaps about 1%. In a stand with different stages of maturity, the light intensity on the forest floor varies (. . .). The higher light intensity on the ground where there are gaps in the cover favor low growth that requires more light ($\triangle \triangle \triangle$). Elsewhere the low growth is shade tolerant ($\blacktriangle \blacktriangle \blacktriangle$). Fruit eating consumers ($\triangledown$) dominate in the areas of mature and aging stands. In the pre-mature stands where leaf production predominates, herbivores are more frequent ($\blacktriangledown$). Source: after Oldeman 1989

in the higher stories of the forest because the limited plant cover on the forest floor, except in periods of regeneration, provides only enough food for a few herbivores. The variety of animals reflects the size of the available living space and the range of local climates and flora. The great height of many trees extends the potential living area for animals vertically so that there exists a complex three dimensionally structured space in which different animal species can settle in one of the various habitats created. The availability of food and the structure of the living space remains unchanged throughout the year. The food supply, a function of the very large primary production of the tropical rainforest, is exceptionally large.

15.5.4
Biomass and primary production

High solar radiation, equable temperature year-round, high humidity and precipitation distributed throughout the year provide a basis for forest growth in the tropical rainforest, a large *biomass* and high *primary production*, despite the widespread limited soil fertility.

Estimates and measurements of the biomass are available from a great many forests. Most range from 300 t to 650 t ha^{-1}. From 75 to 90% of the biomass is above ground and of this, 90% is in the form of wood in living trees, The leaf mass is about 2%, in absolute amount more than in any other ecozone.

Primary production also exceeds that of any other ecozone. Estimates based on climatic parameters and partial measurements of other components such as the litter, have given values that range from 20 to 30 $t\,ha^{-1}\,a^{-1}$. The highest productivity occurs during the rejuvenation phases. Most plants produce only at certain periods of the year, despite conditions that are suitable year-round for primary production. Nevertheless the periods of production are longer than in all other ecozones which may explain the higher production levels of the tropical rainforest.

15.5.5
Animal feed

Because of the great variety of species but relatively low densities of animals, food chains and webs are very complicated. Many animals are difficult to find and knowledge is limited to all but a few animal species or groups. The *zoomass* is very small and the importance, quantitatively, of the consumption on the turnover of energy and matter in the rainforest ecosystem must also be small and characterized by a short cycle between producers and decomposers with almost no intervention from herbivores. It has been suggested that animals in the rainforest have more of a regulatory role on supplies and processes in the forest system than any immediate effect on the turnover.

15.5.6
Litter, the litter layer, decomposition and humus

Since losses from animal feeding are unimportant, the long term delivery of *litter* in the tropical rainforest is about the same as the above ground net primary production, probably between 15 and 25 $t\,ha^{-1}\,a^{-1}$, of which about 5 to 10 $t\,ha^{-1}\,a^{-1}$ is from leaf fall, 80% of the total leaf mass. This is the proportion of leaves renewed each year. The mean life span of the leaves is about 15 months.

Despite the considerable amount of litter supplied, the forest floor is not usually covered by a closed litter layer. The carbon content of the litter is only 1% of the total organic carbon found in the forest ecosystem, compared to 10% in the deciduous forests of the Temperate midlatitudes. The litter mass is so low because the biological and chemical *decomposition* of organic matter in the continuously warm humid conditions that prevail in the trunk area and on the forest floor is very rapid. Leaf litter can be decomposed within a few months (Table 5.3). Depending on the thickness, dead tree wood is completely decomposed in a few, at most, 15 years.

Humus content is therefore low. In the upper soil it is usually between 1 and 3%, 50 to 150 $t\,ha^{-1}$, less than in the soils under temperate forests. Fungi, termites and earthworms are all involved in the decomposition process. The decomposition by bactaria can be limited in acid soils. Many fungi are mycorrhiza and live in symbiosis with higher plants. Termites take care of the

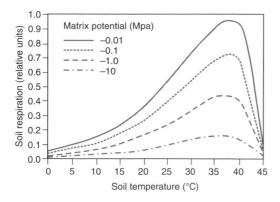

Fig. 15.8. The interdependence of the rate of decomposition (soil respiration), soil temperature and soil moisture. Higher temperatures, particularly in relation to a higher moisture content, and therefore matrix potential, have an accelerating effect on decomposition. Source: Scholes et al.

decomposition of dead wood. The earthworm zoomass is the largest of the decomposers.

15.5.7
Mineral supplies and turnovers

The absolute amount of minerals both in total and for individual minerals per spatial unit are greater and the flows of minerals per time unit and area are higher in the tropical rainforests than in any other forest area. Within the rainforest, the highest values are on soils that are both well supplied with minerals and where concentrations of minerals in the biomass are also larger.

The distribution of the most important mineral supplies and turnovers is shown in Fig. 15.9. It has been suggested that most of the minerals circulating in the rainforest system were in the biomass and not in the soil, as is the case in temperate woodland. Research indicates that this is true only for locations that are very poor in nutrients.

The ability of the rainforest system to keep losses through leaching small is largely the result of the dense root systems in the upper soil, sometimes with roots spread over the soil surface in contact with the litter and also their connection to the mycorrhiza mycelium. In this way, not only the nutrient elements brought in by the precipitation, canopy leaching and stem flow, but also the nutrients in the organic waste are intercepted and supplied to the tree roots directly. Leaching losses that do occur are compensated for by external inputs from precipitation (Table 15.2).

Minerals from the biomass are returned by means of the waste production and canopy leaching. Fire is unimportant. Leaf litter is particularly important in the return of minerals. Although at most half the supply of litter is from leaves, their share of minerals in the litter is far higher than that provided by

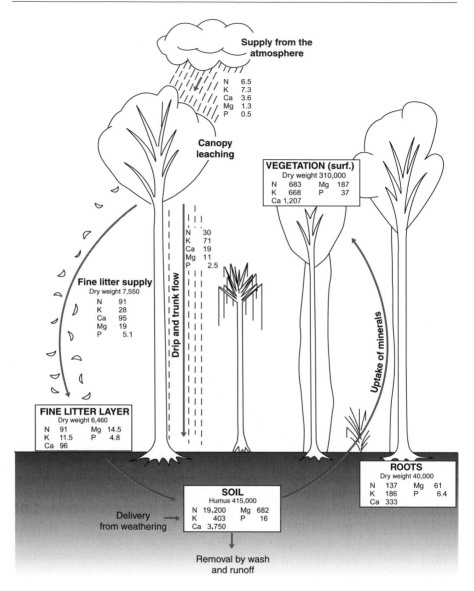

Fig. 15.9. Mineral supply and turnover in an upland rain forest in New Guinea. Supply = kg ha^{-1} Turnover kg ha^{-1} a^{-1}. The relatively rich nutrient supply in and on the soil is related to the high humus content of upland forests where the weight of dead organic soil matter may be greater than the biomass weight. The proportion of minerals in the above and below ground biomass is 29% for calcium and 26% for magnesium, less than half of the content for these minerals in the forest system as a whole, including the soil. The proportion of potassium and phosphorous bound in the biomass is 67% and 68% respectively. Source: Edwards 1982

Table 15.2. Minerals supplied to the soil from rainfall, canopy leaching (drip and stem flow) and litter fall in areas of two tropical rainforests in the Ivory Coast

Research area	Mineral supply to soil	N	K	Ca	Mg	P
Plateau	Total ($kg\,ha^{-1}\,a^{-1}$)	258	85	97	91	9.8
	from (%)					
	– rainfall	9	6	22	4	14
	– canopy leaching	25	61	15	40	4
	– litter fall	66	33	63	56	82
Valley	Total ($kg\,ha^{-1}\,a^{-1}$)	246	264	135	90	24
	from (%)					
	– rainfall	10	2	16	4	6
	– canopy leaching	26	67	21	56	38
	– litter fall	64	31	63	40	56

Source: Bernhard-Reversat 1975

fallen branches and trunks, despite loss from the leaves through leaching and the translocation of nutrient elements into the shoots before leaf drop.

Canopy leaching is of particular importance for potassium supplies. Often more than twice as much potassium is returned to the soil in this way than from litter fall. Potassium, therefore, circulates much faster than calcium for example. The canopy also returns a larger proportion of magnesium and phosphorous. For all other nutrient elements, including nitrogen and calcium, the litter fall is much more important (Table 15.2).

The concept that the mineral turnover in the rainforest ecosystem is a closed system of circulation is only partially correct (Fig. 15.10). The tropical rainforests do, however, have a particularly efficient mechanism to obtain their mineral supplies, indicated by the lack of soluble or suspended load in the streams and also by the fact that soil fertility after forest clearing, and therefore stopping of the return flows, declines rapidly.

15.5.8
Rainforest ecosystem

Figure 15.11 shows the typical available supplies and turnovers in a tropical rainforest system. Characteristic are the great diversity of species and the constant external conditions and internal processes in the system. This means that the accumulation of litter and its decomposition, the translocation of minerals from the leaves before leaf drop and the uptake of minerals by the vegetation from the soil are continuous and optimally synchronized during the entire year. In this way, the system is protected from nutrient shortages and leaching. Agricultural use destroys this balance.

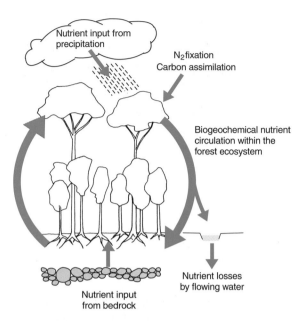

Fig. 15.10. Nutrient supply in a forest ecosytem. Material flows are also received from precipitation and from the bedrock, including seepage and flowing water. Gases are also supplied to the system from nitorgen fixation (N₂) and photosynthesis (carbon dioxide) as well as by the release of nitrogen and phosphorus as a result of decomposition and fire. The amounts supplied from external (thin arrows) are less than those circulating within the system (thick arrows) but are important for the nutrient budget in the forest

15.6
Land use

The tropical rainforests are one of the two remaining forest areas in the world in which agricultural development is still largely limited to areas on the margin of the forest or to island like developments in its interior. The other is the boreal forest (Fig. 6.2). Nevertheless once clearing has begun, these areas increase rapidly in size, as either cutting for timber or for cultivation spread. Fire is also used as a means to clear the forest for agricultural use.

Poor soil quality has limited cultivation in many areas leaving these regions thinly settled (Table 15.3). In general, only *shifting cultivation* is possible if traditional methods are used. Crops such as cassava, taro and yams are grown for a few years in the clearings until a new clearing is prepared by burning the felled branches of trees with their foliage, seldom the tree itself, which provide fertilizer in the form of ash. Harvests in the clearings decline after a few years. Once burnt over, a period of 15 to 30 years is required before a clearing can be successfully cultivated again. Very large areas of land are consumed by this method of agricultural production. Settlements are moved after an even longer

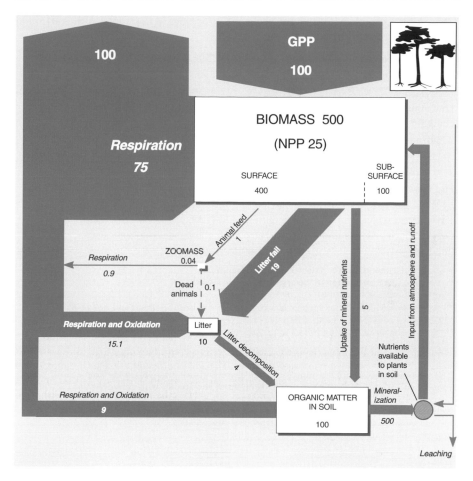

Fig. 15.11. Model of ecosystem of a tropical rain forest. Compared to other zones, the tropical rain forests have 1. a larger biomass and smaller supply of litter and organic matter in the soil, 2. a high turnover rate of energy and minerals and 3. in poor locations, low quantities of exchangeable nutrient elements in the soil. Width of arrows, areas of circle and boxes are approximately in proportion to the volumes involved. Organic substances in ha^{-1} or $ha^{-1} a^{-1}$, minerals in $kg\,ha^{-1}\,a^{-1}$

period when the distance to the new clearing becomes too great. The migration of the families takes place within traditionally defined forest areas.

The decline in productivity the longer an area is cultivated was thought to be caused by a decline in nutrients because of the uptake by the crops and to soil leaching. Research in a tropical forest in southern Venezuela near San Carlos de Rio Negro, indicates, however, that supplies of the most important minerals in the soil do not decline during the normal three year cultivation of an area with manioc (Jordan 1987). It was found that output is reduced, for example,

Table 15.3. Distribution of disadvantageous soil properties in the tropics with year-round rain and the tropics with seasonal rain

Disadvantageous soil characteristics	Humid tropics[f]		Moist savanna		Dry and thorn savanna	
	10^6 ha	%	10^6 ha	%	10^6 ha	%
Low mineral nutrient content[a]	929	(64)	287	(55)	166	(16)
Aluminum toxicity[b]	808	(56)	261	(50)	132	(13)
Acidity without Al toxicity[c]	257	(18)	264	(50)	298	(29)
High phosphate fixation[d]	537	(37)	166	(32)	94	(9)
Low cation exchange capacity[e]	165	(11)	19	(4)	63	(6)
Total area	1444	(100)	525	(100)	1012	(100)

[a] < 10% of usable minerals in the silt and sand fractions
[b] > 60% of the Al saturation in the upper soil layer (1–50 cm)
[c] pH < 5 (low base saturation)
[d] only in soils with high clay content, absent in sandy and loamy Acrisols and Ferralsols
[e] CEC_{eff} < 4 cmol(+)kg^{-1} (with a pH value of 7, this corresponds to 7 cmol(+)kg^{-1}).
  Soils are liable for leaching (which takes place most frequently in Arenosols, Podzols and Acrisols).
[f] The humid tropics cover almost the same area as the Tropics with year-round rain ecozone.

The total area of soils with disdvantageous soil characteristics is highest in the Tropics with year-round rain and next highest in the moist savanna of the Tropics with summer rain. Individual characteristics are, however, frequently interrelated and occur together in many Arenosols, Podzols, Acrisols and Ferralsols, for example, so that even in the humid tropics there are considerable areas of soils without any of these five disadvantageous characteristics.
Source: Sanchez and Logan 1992

if there is an increase in the fixation of phosphates and in aluminum toxicity, both of which increasingly affect plants. An initial rise in the *pH value* from 3.9 to 5.4, which followed the burning of an area and its fertilization with ash, was paralleled by an increase in the available phophates and reduction in toxic aluminum. When the land was cultivated the pH values soon declined to 4.1 and later 3.8 and with this decline phosphate fixation and aluminium toxicity rose again to former high levels. This indicates that the benefit of burning off the land lies in the increase in pH, rather than the release of plant nutrients (Fig. 15.12).

Ecologically, slash and burn cultivation has been a sound system. Yields, even when only for a family's subsistence, are, however, small in relation to the high input of labor. Eventually also, the time span available between the phases of cultivation is too short for the forest to regenerate. This occurs already at a population density of 6 per square kilometer because the area required by a family is very large, with a factor of 10 to 15 (Chap. 14.6).

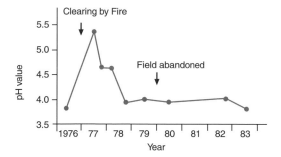

Fig. 15.12. Changes in the pH values in the soil in an Amazonian rain forest after clearing by fire, followed by a three year period of traditional shifting cultivation. Source: Jordan 1987

Slash and burn cultivation can be replaced by modern farming methods. The level of plant nutrients in the soil after clearing is moderately high. Also, the initial decline in humus after clearing levels off and, depending on the type of cultivation or degree of forest regeneration, may even increase again. With the addition of organic matter to the soil in the form of mulch and of lime to raise the pH value, the cation exchange capacity increases. Toxic aluminum is also removed and the availability of phosphates improves. High yields have been achieved in several areas of the humid tropics after the introduction of suitable forms of land use and cultivation, indicating that long term success using such methods is possible.

In Yurimaguas, for example, in Peru in the western Amazon basin the soil is a sandy Acrisol with a high aluminum content and shortages of phosphorous and potassium and most other nutrient elements, together with a pH value of a little over 4. Mean annual precipitation is 2,200 mm. With the addition of fertilizer and lime at a rate of $3\,t\,ha^{-1}$ every three years and suitable rotation

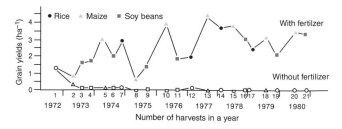

Fig. 15.13. Development of yields on permanently cultivated fields on an Amazonian Acrisol in Yurimaguas, Peru. The upper curve shows yields using fertilizers (initially 80-100-80 kg N-P-K per hectare and later, when the pH reaches 5.5, of 100-26-80 kg N-P-K per hectare and root crop) together with a crop rotation of hill rice, maize and soy beans. In most years two crops were grown, in some three. After 25 harvests the mean yield was $7.8t\ ha^{-1}\ a^{-1}$. Yields did not decrease during the period, in contrast to yields where no fertilizer was applied. After three harvests, yields were almost zero. Source: Jordan 1987, Bandy and Sanchez 1986

of crops, the nutrient content of the soil and the pH value were considerably increased and aluminum content decreased, so that high yields could be maintained (Fig. 15.13).

A large share of the cultivated area in the ecozone is taken up by *tree crops*, more than in any other ecozone. Production is from both small and middle-sized family holdings with several different crops and from large scale *plantations* which tend to specialize on one crop (monoculture). Many plantations process their crop on the plantation, often for the export market, bringing in labor from outside so that income levels are generally higher. Also characteristic of plantations are high capital investment requirements per spatial unit and a dependence on world markets. Their risk is that they have a low production flexibility, especially if trees are grown which take a long time to mature.

The trees, shrubs or lianas in the tropics include rubber, oil and coconut palm, cocoa, spices such as pepper, cinnamon, vanilla, nutmeg, cloves and allspice, as well as coffee and tea. Pineapple and sugar cane are grown in large plantations as field crops and cassava on small farms.

Considerable areas of former forests, especially in South America have been developed as large scale *cattle grazing* enterprises in recent years. These are now, spatially, the largest form of land use in many areas.

The tropics are still cultivated for the most part using traditional farming methods, despite the natural potential of these areas. Lack of capital to invest in the land and a lack of knowledge are the main causes preventing the use of more modern methods. In many cases, the introduction of new forms of cultivation are frustrated because the increased production costs resulting from the purchase of seeds, pecticides and fertilizer necessary for the introduction and maintenance of new forms of cultivation are not covered by an increase in income or profitability.

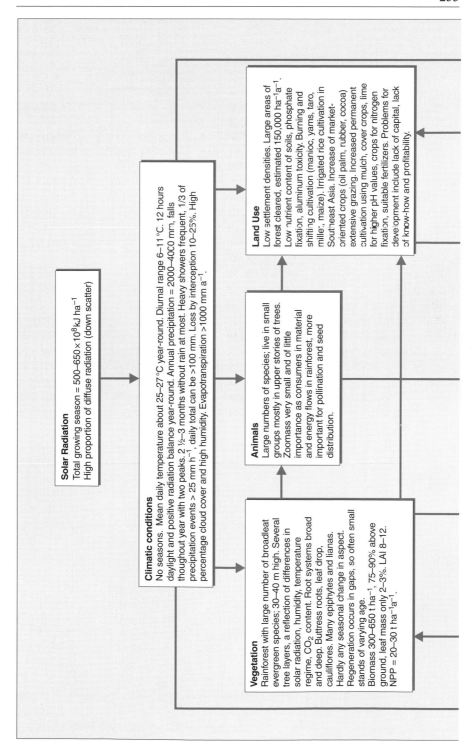

Solar Radiation
Total growing season = 500–650×10^8 kJ ha^{-1}
High proportion of diffuse radiation (down scatter)

Climatic conditions
No seasons. Mean daily temperature about 25–27°C year-round. Diurnal range 6–11°C. 12 hours daylight and positive radiation balance year-round. Annual precipitation = 2000–4000 mm, falls thoughout year with two peaks. 2½–3 months without rain at most. Heavy showers frequent, 1/3 of precipitation events > 25 mm h^{-1}, daily total can be >100 mm. Loss by interception 10–25%. High percentage cloud cover and high humidity. Evapotranspiration >1000 mm a^{-1}.

Vegetation
Rainforest with large number of broadleaf evergreen species; 30–40 m high. Several tree layers, a reflection of differences in solar radiation, humidity, temperature regime, CO$_2$ content. Root systems broad and deep. Buttress roots, leaf drop, cauliflores. Many epiphytes and lianas. Hardly any seasonal change in aspect. Regeneration occurs in gaps, so often small stands of varying age. Biomass 300–650 t ha^{-1}, 75–90% above ground, leaf mass only 2–3%. LAI 8–12. NPP = 20–30 t ha^{-1}a^{-1}.

Animals
Large numbers of species; live in small groups mostly in upper stories of trees. Zoomass very small and of little importance as consumers in material and energy flows in rainforest, more important for pollination and seed distribution.

Land Use
Low settlement densities. Large areas of forest cleared, estimated 150,000 ha^{-1}a^{-1}. Low nutrient content of soils, phosphate fixation, aluminum toxicity. Burning and shifting cultivation (manioc, yams, taro, mille., maize). Irrigated rice cultivation in Southeast Asia. Increase of market-oriented crops (oil palm, rubber, cocoa) extensive grazing. Increased permanent cultivation using mulch, cover crops, lime for higher pH values, crops for nitrogen fixation, suitable fertilizers. Problems for development include lack of capital, lack of know-how and profitability.

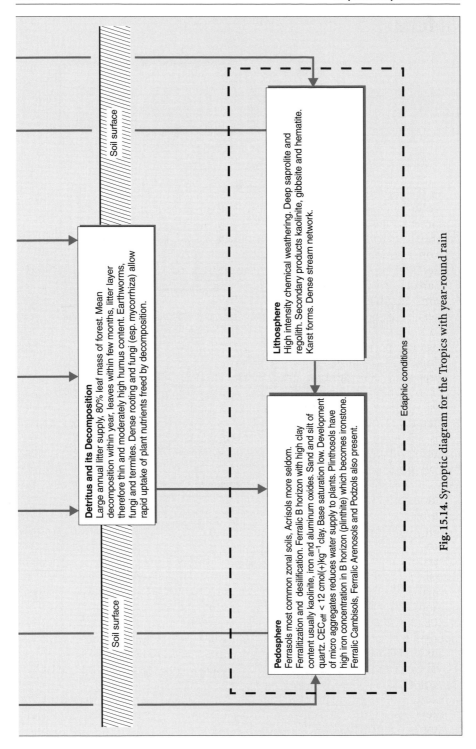

Detritus and its Decomposition
Large annual litter supply, 80% leaf mass of forest. Mean decomposition within year, leaves within few months, litter layer therefore thin and moderately high humus content. Earthworms, fungi and termites. Dense rooting and fungi (esp. mycorrhiza) allow rapid uptake of plant nutrients freed by decomposition.

Soil surface

Lithosphere
High intensity chemical weathering. Deep saprolite and regolith. Secondary products kaolinite, gibbsite and hematite. Karst forms. Dense stream network.

Pedosphere
Ferrasols most common zonal soils, Acrisols more seldom. Ferrallitization and desilification. Ferralic B horizon with high clay content usually kaolinite, iron and aluminum oxides. Sand and silt of quartz. $CEC_{eff} < 12$ cmol(+)kg^{-1} clay. Base saturation low. Development of micro aggregates reduces water supply to plants. Plinthosols have high iron concentration in B horizon (plinthite) which becomes ironstone. Ferralic Cambisols, Ferralic Arenosols and Podzols also present.

Edaphic conditions

Fig. 15.14. Synoptic diagram for the Tropics with year-round rain

References

Aber JD, Melillo JM, Nadelhoffer KJ, Pastor J, Boone RD (1991): Factors controlling nitrogen cycling and nitrogen saturation in northern temperate forest ecosystems. Ecol. Applications 1, 303–315.

Achtnich W, Lüken H (1986): Bewässerungslandbau in den Tropen und Subtropen. In: Rehm, 285–342.

Ahnert F (1987): An approach to the identification of morphoclimates. In: Gardiner V (ed.): International geomorphology, Part II. John Wiley and Sons, Chichester, 159–188.

Ahnert F (2003): Einführung in die Geomorphologie. Ulmer, Stuttgart (3rd ed.).

Aleksandrova VD (1988): Vegetation of the Soviet polar desert. Cambridge Univ. Press, Cambridge.

Andreae B (1983): Agrargeographie. De Gruyter, Berlin (2nd ed.).

Aoki M, Yabuki K, Koyama H (1975): Micrometeorology and assessment of primary production of a tropical rainforest in West Malaysia. J. Agric. Met. (Japan) 31:115–124.

Apps MJ, Kurz WA, Luxmoore RJ, Nilsson LO, Sedjo RA, Schmidt R, Simpson LG, Vinson TS (1993): Boreal forests and tundra. Water, Air and Soil Pollution 70:39–53.

Apps MJ, Price DT (eds) (1996): Forest ecosystems, forest management and the global carbon cycle. NATO ASI Ser/G 40. Springer, Berlin.

Archibold OW (1995): Ecology of world vegetation. Chapman and Hall, London.

Arianoutsou M, Groves RH (eds) (1994): Plant-animal interactions in mediterranean-type ecosystems. Kluwer, Dordrecht.

Arnold A (1997): Allgemeine Agrargeographie. Klett-Perthes, Gotha.

Arroyo MTK, Zedler PH, Fox MD (eds) (1995): Ecology and biography of mediterranean ecosystems in Chile, California and Australia. Ecol. Studies 108. Springer, Berlin.

Bandy DE, Sanchez PA (1986): Post-clearing soil management alternatives for sustained production in the Amazon. In: Lal et al., 347–361.

Batzli GO (1981): Populations and energetics of small mammals in the tundra ecosystem. In: Bliss et al., 377–396.

Bazilevich NI, Rodin LY (1971): Geographical regularity in productivity and the circulation of chemical elements in the earth's main vegetation types. Soviet Geography, New York, 24–53.

Beaumont P (1993): Drylands – environmental management and development. Routledge, London (2nd ed.).

Bernhard-Reversat F (1975): Nutrients in throughfall and their quantitative importance in rain forest mineral cycles. In: Golley and Medina, 153–159.

Besler H (1992): Geomorphologie der ariden Gebiete. Wiss. Buchges., Darmstadt.

Bick H (1993): Ökologie. Fischer, Stuttgart (2nd ed.).

Bird MJ, Chivas AR, Head J (1996): A latitudinal gradient in carbon turnover times in forest soils. Nature 381:143–146.

Bliss LC (1997): Arctic ecosystems of North America. In: Wielgolaski, 551–683.

Bliss LC, Heal OW, Moore JJ (eds) (1981): Tundra ecosystems: a comparative analysis. Intern. Biol. Progr. 25. Cambridge University Press, Cambridge.

Blume H (1991): Das Relief der Erde, Ein Bildatlas. Enke, Stuttgart.

Blume H-P, Berkowicz SM (eds) (1995): Arid ecosystems. Adv. Geoecology 28.

Blume H-P, Felix-Henningsen P, Fischer WR, Frede H-G, Horn R, Stahr K (eds) (published in series since 1996): Handbuch der Bodenkunde. Ecomed, Landsberg.

Blümel WD (1999): Physische Geographie der Polargebiete. Teubner, Stuttgart.

Blüthgen J, Weischet W (1980): Allgemeine Klimageographie. De Gruyter, Berlin (3rd ed).

Bonan GB, Chapin III FS, Thompson SL (1995): Boreal forest and tundra ecosystems as components of the climate system. Climatic change 29:145–167.

Boose ER, Foster DR, Fluet M (1994): Hurricane impacts to tropical and temperate forest landscapes. Ecol. Monographs 64:369–400.

Booysen P de V, Tainton NM (eds) (1984): Ecological effects of fire in South African ecosystems. Ecol Studies 48. Springer, Berlin.

Bourliere F (ed) (1983): Tropical savannas. Ecosystems of the World 13, Elsevier, Amsterdam.

Box EO, Peet RK, Masuzawa T, Yamada I, Fujiwara K, Maycock PF (eds) (1995): Vegetation science in forestry. Kluwer Acad. Publ., Dordrecht.

Breymeyer AI, Van Dyne GM (eds) (1980): Grasslands, systems analysis and man. Intern. Biol. Progr. 19. Cambridge University Press, Cambridge.

Bridges EM, Batjes NH, Nachtergaele FO (eds) (1998): World reference base for soil resources: atlas. Acco, Leuven.

Bruenig EF (1996): Conservation and management of tropical rainforest – an integrated approach to sustainability. Centre Agric. Biosci. (Cab) Intern., Wallingford.

Büdel, J (1981): Klima-Geomorphologie. Borntraeger, Berlin (2nd ed).

Bullock SH, Mooney HA, Medina E (eds) (1995): Seasonally dry tropical forests. Cambridge Univ. Press, Cambridge.

Burke IC, Yonker CM, Parton WJ, Cole CV, Flach K, Schimel DS (1989): Texture, climate and cultivation effects on soil organic matter content in U.S. grassland soils. Soil Sci. Soc. Am. J. 53:803–805.

Busche D (1998): Die zentrale Sahara: Oberflächenformen in Wandel. Justus Perthes Verlag, Gotha.

Butzer KW (1976): Geomorphology from the Earth. Harper and Row, New York.

Cernusca A (1975): Eine neue Ausbildungsmethode für Umweltforschung. Umschau 75:242–245.

Chabot BF, Mooney HA (1985): Physiological ecology of North American plant communities. Chapman and Hall, New York.

Chadwick AC, Sutton SL (eds) (1984): Tropical rain-forest: The Leeds Symposium, Philosophical and Library Society, Leeds.

Chapin III FS, Jefferies RL, Reynolds JF, Shaver GR, Svoboda J (eds) (1992): Arctic ecosystems in a changing climate: an ecophysiological perspective. Academic Press, New York.

Chapin III FS, Körner C (eds) (1995): Arctic and alpine biodiversity. Ecol. Studies 113. Springer, Berlin.

Chernov YI, Matveyeva NV (1997): Arctic ecosystems in Russia. In: Wielgolaski, 361–507.

Chorley RJ, Schumm SA, Sugden DE (1984): Geomorphology. Methuen, New York.

Cloudsley-Thompson JL (1996): Biotic interactions in arid lands. Springer, Berlin.

Cole MM (1986): The savannas: biogeography and geobotany. Academic Press, London.

Cole DW, Rapp M (1981): Elemental cycling in forest ecosystems. In: Reichle, 341–409.

Coleman SM, Dethier DP (eds) (1986): Rates of chemical weathering of rocks and minerals. Academic Press, Orlando.

Commission of Euroean Community (1985): Soil Map of the European Communities 1:1,000,000. Luxemburg.

Conrad CE, Oechel WC (eds) (1982): Dynamics and management of mediterranean-type ecosystems. Pacific Southwest Forest and Range Experiment Station, Berkeley.

Coupland RT (ed) (1979): Grassland ecosystems of the world: analysis of grasslands and their uses. Intern. Biol. Progr. 18. Cambridge University Press, Cambridge.

Coupland RT, Van Dyne GM (1979): Natural temperate grasslands: systems synthesis. In: Coupland, 97–106.

Coupland RT (ed) (1992, 1993): Natural grasslands. Ecosystems of the World 8A and 8B. Elsevier, Amsterdam.

Cramer WP, Solomon AM (1993): Climatic classification and future global redistribution of agricultural land. Climate Research 3:97–110.

Davis GW, Richardson DM (eds) (1995): Mediterranean-type ecosystems: the function of biodiversity. Ecol. Studies 109. Springer, Berlin.

Day AD, Ludeke KL (1993): Plant nutrients in desert environments. Springer, Berlin.

Day JA (ed) (1983): Mineral nutrients in mediterranean ecosystems S. Afri. Nat. Sci. Prog. Rep. 71. CSIR, Pretoria.

De Angelis DL, Gardner RH, Shugart HH (1981): Productivity of forest ecosystems studied during the IBP: the woodlands data set. In: Reichle, 567–672.

Deckers JA, Nachtergaele FO, Spaargaren OC (eds) (1998): World reference base for soil resources: introduction. Acco, Leuven.

De Jong B (1973): Net radiation recieved by a horizontal surface at the earth. Delft University Press, Delft.

Dell B, Hopkins AJM, Lamont BB (eds) (1986): Resilience in mediterranean-type ecosystems. Tasks Veg. Sci. 16. Dr. W. Junk, The Hague.

Di Castri F, Godall DW, Specht RL (eds) (1981): Mediterranean-type shrublands. Ecosystems of the World 11. Elsevier, Amsterdam.

Dickinson RE (ed) (1987): The geophysiology of Amazonia. Vegetation and climate interactions. John Wiley and Sons, New York.

Doppler W (1998): Landwirschaftliche Betriebssysteme in den Tropen und Subtropen. Ulmer, Stuttgart (2nd ed.).

Driessen PM, Dudal R (eds) (1991): The major soils of the world. Agric. Univ. Wageningen and Kath. Univ. Leuven.

Duvigneaud P (ed) (1971): Productivity of forest ecosystems. UNESCO, Paris.

Eden MJ (1990): Ecology and Land management in Amazonia. Belhaven, London.

Edwards PJ (1982): Studies of mineral cycling in a montane rain forest in New Guinea. Rates of cycling in throughfall and litter fall. J. Ecol 70:807–827.

Eitel B (1999): Bodengeographie. Das Geographische Seminar. Westermann, Braunschweig.

Elias P, Kratochvilova I, Janous D, Marek M, Masarovicova E (1989): Stand microclimate and physiological activity of tree leaves in an oak-hornbeam forest. Trees 4:227–233.

Ellenberg H, Mayer R, Schauermann J (eds) (1986): Ökosystemforschung. Ergebnisse des Sollingprojekts 1966–1986. Ulmer, Stuttgart.

Elliot-Fisk DL (1989): The boreal forest. In: Barbour MG, Billings WD (eds) (1989): North American terrestrial vegetation. Cambridge, 33–62.

Esser G, Overdieck D (eds) (1991): Modern ecology: basic and applied aspects. Elsevier, Amsterdam.

Evenari M, Noy-Meir I, Goodall DW (eds) (1985, 1986): Hot desert and arid shrublands. Ecosystems of the World 12A and 12B. Elsevier, Amsterdam.

Falinski JB (1986): Vegetation dynamics in temperate lowland primeval forests. Ecological studies in Bialowieza forest. Geobotany 8. Dr. W. Junk, Dordrecht.

FAO-UNESCO (1974–1981): Soil Map of the World, Vol. I–X and 18 maps 1:5 million. UNESCO, Paris.

FAO (1988, 2nd ed. 1990): Revised legend of the FAO-UNESCO Soil Map of the World. World Soil Resources Rep. 60, Rome.

FAO (1994): Soil map of the world – revised legend with corrections. ISRIC Technical Paper 20, Wageningen.

Finkel HJ (1986): Semiarid soil and water conservation. CRC Press, Boca Raton.

Fischer A (1995): Forstliche Vegetationskunde. Pareys Studientexte 82. Blackwell-Wiss. Verlag, Berlin.

French HM (1976): The periglacial environment. Longman, London.

French HM (1981): Permafrost and ground ice. In: Gregory KJ, Walling DE (eds): Man and environmental processes. Butterworth, London, 144–162.

French NR (ed) (1979): Perspectives in grassland ecology. Ecol. Studies 32. Springer, Berlin.

Frey W, Lösch R (2004): Lehrbuch der Geobotanik. Fischer, Stuttgart (2nd ed.).

Frost PGH, Menaut J-C, Walker B, Medina E, Solbrig OT, Swift M (eds) (1986): Responses of savannas to stress and disturbance. Biol. International (IUBS News Magazine), Special Issue 10.

Furley PA, Proctor J, Ratter JA (eds) (1992): Nature and dynamics of forest-savanna boundaries. Chapman and Hall, London.

Ganssen R (1965): Grundsätze der Bodenbildung. Bibliographisches Institut, Mannheim.

Gibson AC (1996): Structure-function relations of warm desert plants. Springer, Berlin.

Giessner K (1988): Die subtropisch-randtropische Trockenzone. Globale Verbreitung, innere Differenzierung, geoökologische Typisierung und Bewertung. Geoökodynamik 9:135–183.

Golley FB, Medina E (eds) (1975): Tropical ecological systems: trends in terrestrial and aquatic research. Ecol. Studies 11. Springer, Berlin.

Golley FB (ed) (1983): Tropical rainforest ecosystems. Ecosystems of the World 14A. Elsevier, Amsterdam.

Goodall DW, Perry RA (eds) (1979, 1981): Arid-land ecosystems: structure, functioning and management. Intern.Biol. Prog. 16 and 17. Cambridge University Press, Cambridge.

Gore AJP (ed) (1983): Mires: swamp, bog, fen, more. Ecosystems of the World 4A and 4B. Elsevier, Amsterdam.

Gorham E (1991): Northern peatlands: role in the carbon cycle and probable responses to climatic warming. Ecol. Applications 1:182–195.

Grigg DB (1974): The agricultural systems of the world. Cambridge University Press, Cambridge.

Häckel H (1999): Meteorologie. Ulmer, Stuttgart (4th ed.).

Hagedorn J, Poser H (1974): Räumliche Ordnung der rezenten geomorphologischen Prozesse and Prozesskombinationen auf der Erde. In: Poser H (ed): Geomorphologische Prozesse und Prozesskombinationen in der Gegenwart unter verschiedenen Klimabedingungen. Vandenhoeck and Ruprecht, Göttingen, 426–439.

Hartzler SA, Huo Y-Q (1995): Comparison of the vegetation of subtropical China with that of the corresponding region of the USA. In: Box et al., 105–123.

Hegarty EE (1991): Leaf litter production by lianes and trees in a sub-tropical Australian rain forest. J. Trop. Ecol. 7:201–214.

Henning D (1989): Atlas of the surface heat balance of the continents. Borntraeger. Berlin.

Henning I (1994): Hydroklima und Klimavegetation der Kontinente. Münsterische Geog. Arb. 37. Münster.

Henry GHR, Gunn A (1991): Recovery of tundra vegetation after overgrazig by Caribou in arctic Canada. Arctic 44:38–42.

Herrmann R (1977): Einführung in die Hydrologie. Teubner, Stuttgart.

Hobbs RJ (ed) (1992): Biodiversity of Mediterranean ecosystems in Australia. Surrey Beatty, Chipping Norton.

Hodgkinson K (ed) (1997): Landscape ecology, function and management: principles from Australia's rangelands. CSIRO, Australia.

Hofmeister B (1985): Die gemäßigten Breiten. Geographisches Seminar Zonal. Westermann, Braunschweig.

Holdridge LR (1947): Determination of world plant formations from simple climatic data. Science 105:367–368.

Holm-Nielsen LB, Nielsen IC, Balslev H (eds) (1989): Tropical forests – botanical dynamics, speciation and diversity. Acad. Press, London.

Holtmeier FK (1985): Die klimatische Waldgrenze – Linie oder Übergangssaum (Ökoton)? Erdkunde 39:271–285.

Hübl E (1988) Lorbeerwälder und Hartlaubwälder (Ostasien, Mediterraneis and Makronesien). Düsseldorfer Geobot. Kolloq. 5:3–26.

Hueck K (1966): Die Wälder Südamerikas. Fischer, Stuttgart.

Huntley BJ, Walker BH (eds) (1982): Ecology of tropical savannas. Ecol. Studies 42. Springer, Berlin.

Hupfer P, Kuttler W (1998): Heyer – Witterung und Klima. Teubner, Leipzig (10th ed.).

Hustich I (1966): On the forest tundra and the northern tree-lines. Ann. Univ. Turku A II 36:7–47.

ISSS-ISRIC-FAO (1988): World reference base for soil resources. World Soil Resources Rep. 84, Rome.

Ives JD, Barry RG (eds) (1974): Arctic and alpine environments. Methuen, London.

Jacobs M (1988): The tropical rain forest. A first encounter. Springer, Berlin

Jahn R (1997): Bodenlandschaften subtropischer mediterraner Zonen. In: Blume et al.

Jakucs P (ed) (1985): Ecology of an oak forest in Hungary. Akadiado, Budapest.

Jatzold R (1984): Das System der agro-ökologischen Zonen der Tropen als angewandte Klimageographie mit einem Beispiel aus Kenia. 44. Dt. Geogr. Tag. Steiner, Stuttgart.

Jatzold R (1984): Steppengebiete der Erde. Praxis Geogr. 14:10–15.

Johansson D (1974): Ecology of vascular epiphytes in West African rain forest. Acta Phytogeogr. Suecica 59.

Johnson DW, Lindberg SE (eds) (1992): Atmospheric deposition and forest nutrient cycling; a synthesis of the integrated forest study. Ecol. Studies 91. Springer, Berlin.

Jordan CF (1985): Nutrient cycling in tropical forest ecosystems. John Wiley and Sons, Chichester.

Jordan CF (ed) (1987): Amazonian rain forests. Ecol. Studies 60. Springer, Berlin.

Joss PJ, Lynch PW, Williams OB (eds) (1986): Rangelands: A resource under siege. Austr. Acad. Sci., Canberra and Cambridge University Press, Cambridge.

Karte J (1979): Räumliche Abgrenzung und regionale Differenzierung des Periglaziärs. Bochumer Geogr. Arb. 35. Schoeningh, Paderborn.

Kasischke ES, Christensen Jr NL, Stocks BJ (1995): Fire, global warming and the carbon balance of boreal forests. Ecological Applications 5:437–451.

Kasischke ES, Stocks B (eds) (2000): Fire, climate change and carbon cycling in the boreal forest. Ecol. Studies 138. Springer, Berlin.

Kato R, Tadaki Y, Ogawa H (1978): Plant biomass and growth increment studies in Pasoh Forest. Malaysian Nat. J. 30:211–224.

Kauffman S, Sombroek W, Mantel S (1998): Soils of rain forests; characterization and major constraints of dominant forest soils in the humid tropics. In: Schulte and Ruhiyat, 9–20.

Keller R (1961): Gewässer und Wasserhaushalt des Festlands. Teubner, Berlin.

Kellman M, Tackaberry R (1997): Tropical environments; the functioning and management of tropical ecosystems. Routledge, London.

Kira T, Shidei T (1967): Primary production and turnover of organic matter in different ecosystems of the Western Pacific. Jap. J. Ecol. 17:70–87.

Kira T, Ono Y, Hosokawa T (eds) (1978): Biological production in a warm-temperate evergreen oak forest of Japan. JIBP synthesis 18. Tokyo Press, Tokyo.

Kira T (1995): Forest ecosystems of east and south-east Asia in a global perspective. In: Box et al., 1–21.

Klink H-J (1996): Vegetationsgeographie. Das geographische Seminar. Westermann, Braunschweig (2nd ed.).

Kloft W, Gruschwitz M (1988): Ökologie der Tiere. Ulmer, Stutgart (2nd ed.).

Klötzli F (1993): Ökosysteme. Fischer, Stuttgart (3rd ed.).

Kolchugina TP, Vinson TS (1993): Climate warming and the carbon cycle in the permafrost zone of the former Soviet Union. Permafrost and Periglacial Processes 4:149–163.

Kratochwil A, Schwabe A (2001): Ökologie der Lebensgemeinschaften. Ulmer, Stuttgart.

Kreeb KH (1983): Vegetationskunde. Ulmer, Stuttgart.

Kreeb KH (1990): Methoden zur Pflanzenökologie und Bioindikation. Fischer, Jena.

Kruger FJ, Mitchell DT, Jarvis JUM (eds) (1983): Mediterranean-type ecosystems. Ecol. Studies 43. Springer, Berlin.

Kuntze H, Roeschmann G, Schwerdtfeger G (1994): Bodenkunde. Ulmer, Stuttgart (5th ed.).

Kuttler W (ed) (1995): Handbuch zur Ökologie. Analytica, Berlin (2nd ed.).

Lal R (1987): Tropical ecology and physical edaphology. John Wiley and Sons, Chichester.

Lal R, Sanchez PA, Cummings RW Jr (eds) (1986): Land clearing and development in the tropics. Balkema, Rotterdam.

Lal R, Sanchez PA (eds) (1992): Myths and science of soils of the tropics. Soil Science Soc. America Spec. Publ. 29.

Lamotte M, Bourliere F (1983): Energy flow and nutrient cycling in tropical savannas. In: Bourliere, 583–603.

Landsberg HE (ed) (1970–1986): World survey of climatology. Vols. 1–15. Elsevier, Amsterdam.

Larcher W, Bauer H (1981): Ecological significance of resistance to low temperature. In: Lange OL, Nobel PS, Osmond CB, Zeigler H (eds), Encyclopedia of pant physiology 12A. Springer, Berlin, 403–437.

Larcher W (2001): Ökophysiologie der Pflanzen. Ulmer, Stuttgart (6th ed.).

Larsen JA (1980): The boreal ecosystems. Academic Press, New York.

Larsen JA (1982): Ecology of northern lowland bogs and conifer forests. Academic Press, New York.

Larsen JA (1989): The northern forest border in Canada and Alaska. Ecol. Studies 70. Springer, Berlin.

Lauenroth WK, Sala OE (1992): Long-term forage production of North American shortgrass steppe. Ecol. Applications 2:397–403.

Lauer W, Rafiqpoor MD, Frankenberg P (1996): Die Klimate der Erde. Erdkunde 50:275–300.

Lauer W (1999): Klimatologie. Das geographische Seminar. Westermann, Braunschweig (3rd ed.).

Le Houerou HN (1981): Impact of man and his animals on mediterranean vegetation. In: Di Castri et al., 479–521.

Le Houerou HN, Bingham RL, Skerbeg W (1988): Relationship between the variability of annual rainfall and the variability of primary production. J. Arid Environment 15:1–18.

Le Houerou HN (1989): The grazing land ecosystems of the African Sahel. Ecol. Studies 75. Springer, Berlin.

Leigh EG Jr, Rand AS, Windsor DM (eds) (1996): The ecology of a tropical forest. Smithsonian Institution, Washington D.C. (2nd ed.).

Lerch G (1991): Pflanzenökologie. Akademie, Berlin.

Lieth H (1964): Versuch einer kartographischen Darstellung der Produktivität der Pflanzendecke auf der Erde. Geogr. Taschenbuch 1964/65. Steiner, Wiesbaden, 72–80.

Lieth H, Whittaker RH (eds) (1975): Primary productivity of the biosphere. Ecol. Studies 14. Springer, Berlin.

Lieth H, Werger MJA (eds) (1989): Tropical rain forest ecosystems. Ecosystems of the World 14B. Elsevier, Amsterdam.

Likens GE, Bormann FH (1995): Biogeochemistry of a forested ecosystem. Springer, New York.

Long SP, Jones MB, Roberts MJ (eds) (1992): Primary productivity of grass ecosystems of the tropics and subtropics. Chapman and Hall, London.

Louis H, Fischer K (1979): Allgemeine Geomorphologie. De Gruyter, Berlin (4th ed.).

Louw GN, Seely MK (1982): Ecology of desert organisms. Longman, London.

Lovegrove B (1993): The living deserts of southern Africa. Fernwood Press, Vlaeberg (South Africa).

Ludwig JA, Tongway D, Freudenberger D, Noble J, Hodgkinson K (1997): Landscape ecology, function and management: principles from Australia's rangelands. CSIRO, Australia.

Lugo AE, Lowe C (eds) (1995): Tropical forests – management and ecology. Ecol. Studies 112. Springer, Berlin.

Lyr H, Fiedler H-J, Tranquillini W (1992): Phyiologie und Ökologie der Gehölze. Fischer, Jena.

Mackay JR (1972): The world of underground ice. Ann. Ass. Am. Geogr. 62:1–22.

Mainguet M (1991): Desertfication. Natural background and human mismanagement. Springer, Berlin.

Margaris NS (1981): Adaptive strategies in plants dominating Mediterranean-type ecosystems. In: Di Castri et al., 309–315.

Margaris NS, Mooney HA (eds) (1981): Components of productivity of Mediterranean climate regions. Tasks Veg. Sci 4. Dr. W. Junk, The Hague.

Markov VD, Khoroshev PI (1986): The peat resources of the USSR and prospects for their utilization. Int. Peat J. 1:41–47.

Marschner H (1990): Mineral nutrition of higher plants. Acad. Press, London (4th ed.).

Martikainen PJ, Nykänen H, Alm J, Silvola J (1995): Change in fluxes of carbon dioxide, methane and nitrous oxide due to forest drainage of mire sites of different trophy. Plant and Soil, 168/169:571–577.

McCloskey JM, Spalding H (1989): A reconnaissance-level inventory of the amount of wilderness remaining in the world. Ambio 18:221–227.

McNaughton SJ, Sala OE, Oesterheld M (1993): Comparative ecology of African and South American arid to subhumid ecosystems. In: Goldblatt P (ed): Biological relationships between Africa and South America. Yale University Press, New Haven, 548–567.

Medina E (1987): Requirements, conservation and cycles of nurients in the herbaceous layer. In: Walker, 39–65.

Medina E, Mooney HA, Vazques-Yanes C (eds) (1984): Physiological ecology of plants of the wet tropics. Tasks Veg. Sci. 12. Dr. W. Junk, The Hague.

Meentemeyer V, Gardner J, Box EO (1985): World patterns and amount of detrital soil carbon. Earth Surf. Processes and Landforms 10:557–567.

Meigs P (1953): World distribution of arid and semi-arid homoclimates. In: UNESCO: Review of Research on Arid Zone Hydrology. UNESCO, Paris, 203–209.

Meurk CD (1995): Evergreen broadleaved forests of New Zealand and their bioclimatic definition. In: Box et al., 151–197.

Mitchell DT, Coley PGF, Webb S, Allsopp N (1986): Litterfall and decomposition processes in the coastal fynbos vegetation, South Western Cape, South Africa. J. Ecol. 74:977–993.

Monk CD, Day FP Jr (1988): Biomass, primary production, and selected nutrient budgets for an undisturbed watershed. In: Swank and Crossley, 151–159.

Mooney HA (1981): Primary production in mediterranean-climate regions. In: Di Castri et al., 249–255.

Mooney HA, Miller PC (1985): Chapparral. In: Chabot and Mooney, 213–231.

Mooney HA (1988): Lessons from mediterranean-climate regions. In: Wilson EO, Peter F (eds): Biodiversity. Nat. Acad. Press, Washington DC, 157–165.

Moreno JM, Oechel WC (eds) (1994): The role of fire in mediterranean-type ecosystems. Ecol. Studies 107. Springer, Berlin.

Moreno JM, Oechel WC (eds) (1995): Global change and mediterranean-type ecosystems. Ecol. Studies 117. Springer, Berlin.

Müller MJ (1996): Handbuch ausgewählter Klimastationen der Erde. For-schungsstelle Bodenerosion Univ. Trier (5th ed.).

Müller-Hohenstein K (1993): Auf dem Weg zu einem neuen Verständnis von Desertifikation – Überlegungen aus der Sicht einer praxisorientierten Geo-botanik. Phytocoenologia 23:499–518.

Nadelhoffer KJ, Giblin AE, Shaver GR, Linkins AE (1992): Microbial processes and plant nutrient availability in arctic soils. In: Chapin III et al., 281–300.

Norman MJT, Pearson CJ, Searle PGE (1995): The ecology of tropical food crops. Cambridge Univ. Press, Cambridge (2nd ed.).

Oechel WC, Callaghan T, Gilmanov T, Holten JI, Maxwell B, Molau U, Svein-björnsson B (eds) (1997): Global change and arctic terrestrial ecosystems. Ecol. Studies 124. Springer, Berlin.

Ohiagu CE, Wood TG (1979): Grass production and decomposition in southern New Guinea savanna, Nigeria. Oecologia 40:155–165.

Ohmura A (1984): Comparative energy balance for arctic tundra, sea surface, glaciers and boreal forests. GeoJournal 8:221–228.

Oldeman RAA (1989): Dynamics in tropical rain forests. In: Holm-Nielsen et al., 3–21.

Ollier CD (1984): Weathering. Longman, Edinburgh (2nd ed.).

O'Neill RV, De Angelis DL (1981): Comparative Productivity and biomass relations of forest ecosystems. In: Reichle, 411–450.

Olson DF (1983): Temperate broad-leaved evergree forests of the southeastern North America. In: Ovington, 103–105.

Ovington JD (ed) (1983): Temperate broad-leaved evergreen forests. Ecosys-tems of the World 10. Elsevier, Amsterdam.

Ovington JD, Pryor LD (eds) (1983): Temperate broad-leaved evergreen forests of Australia. In: Ovington, 73–101.

Paavilainen E, Päivänen J (1995): Peatland forestry. Ecol. Studies 111. Springer, Berlin.

Peet RK (1981): Changes in biomass and production during secondary forest succession. In: West DC, Shugart HH, Botkin DB (eds): Forest Succession. Springer, New York, 324–388.

Persson T (ed) (1980): Structure and function of northern coniferous forests – an ecosystem study. Ecol. Bull. 32. Swedish Nat. Sci. Res. Council, Stock-holm.

Pieri CJMG (1992): Fertility of soils. A future for farming in the West African Savannah. Springer Ser. Phys. Environment 10. Springer, Berlin.

Polunin N (ed): (1986): Ecosystem theory and application. John Wiley and Sons, Chichester.

Pomeroy LR, Alberts JJ (eds) (1988): Concepts of ecosystem ecology. Ecol. Studies 67. Springer, Berlin.

Post WM, Emanuel WR, Zinke PJ, Stangenberger AG (1982): Soil carbon pools and world life zones. Nature 298:156–159.

Potter CS, Ragsdale HL, Swank WT (1991): Atmospheric deposition and foliar leaching in a regenerating southern Appalachian forest canopy. J. Ecol. 79:97–115.

Price DT, Apps MJ (1995): The boreal forest transect case studies: global change effects on ecosystem processes and carbon dynamics in boreal Canada. Water, Air and Soil Pollution 82:203–214.

Proctor J (ed) (1989): Mineral nutrients in tropical forest and savanna ecosystems. Spec. Publ. Brit. Ecol. Soc. 9. Blackwell, Oxford.

Prowse TD (1994): Environmental significance of ice to streamflow in cold regions. Freshwater Biology 32:241–259.

Pye K (1987): Aeolian dust and dust deposits. Academic Press, London.

Quezel P (1981): Floristic composition and phytosociological structure of sclerophyllous matorral around the Mediterranean. In: Di Castri et al., 107–121.

Rambal S (1994): Fire and water yield: a survey and predictions for global change. In: Moreno and Oechel, 96–116.

Reading AJ, Thompson RD, Millington AC (1995): Humid tropical environments. Blackwell, Oxford.

Rehm S (ed) (1986): Grundlagen des Pflanzenbaues in den Tropen und Subtropen. Hdb. der Landwirtschaft und Ernährung in den Enwicklungsländern 3. Ulmer, Stuttgart (2nd ed.).

Reichle DE (1970): Temperate forest ecosystems. Ecol. Studies 1. Springer, Berlin.

Reichle DE, Franklin JF, Goodall DW (eds) (1975): Productivity of world ecosystems. Nat. Acad. Sci. Seattle, Washington.

Reichle DE (ed) (1981): Dynamic properties of forest ecosytems. International Biol. Progr. 23. Cambridge Univ. Press, Cambridge.

Remmert H (1980): Arctic animal ecology. Springer, Berlin.

Remmert H (ed) (1991): The mosaic-cycle concept of ecosystems. Ecol. Studies 85. Springer, Berlin.

Remmert H (1992): Ökologie. Springer, Berlin (5th ed.).

Richards PW (1996): The tropical rain forest: an ecological study. Cambridge Univ. Press, Cambridge (6th ed.).

Richter M (1997): Allgemeine Pflanzengeographie, Teubner, Stuttgart.

Richter M (2001): Vegetationszonen der Erde. Klett-Perthes, Gotha.

Ricklefs RE (1990): Ecology. Freeman and Company, New York (3rd ed.).

Risser PG, Goodall DW, Perry RA, Howes KMW (eds) (1981): The true prairie ecosystem. US/IBP Synthesis Ser. 16. Dowden, Hutchinson and Ross, Stroudsburg.

Risser PG (1988): Abiotic controls on primary productivity and nutrient cycles in North American grasslands. In: Pomeroy and Alberts, 115–129.

Roda F, Retana J, Gracia CA, Bellot J (eds) (1999): Ecology of Mediterranean evergreen oak forests. Ecol. Studies 137. Springer, Berlin.

Rohdenburg H (1971): Einführung in die klimagenetische Geomorphologie. Lenz, Giessen.

Röhrig E, Ulrich B (eds): (1991): Temperate deciduous forests. Ecosystems of the World 7. Elsevier, Amsterdam.

Rosenzweig ML (1968): Net primary productivity of terrestrial communities: prediction from climatological data. Am. Naturalist 102:67–74.

Rother K (1984): Mediterrane Subtropen. Geographisches Seminar Zonal. Westermann, Braunschweig.

Rouse WR (1981): Man-modified climates. In: Gregory KJ, Walling DE: Man and environmental processes. Butterworths, London, 38–54.

Rundel PW, Parsons DJ (1984): Post-fire uptake of nutrients by diverse ephemeral herbs in chamise chaparral. Oecologia 61:285–288.

Rundel PW, Montenegro G, Jaksic FM (eds) (1998): Landscape disturbance and biodiversity in mediterranean-type ecosystems. Ecol. Stud. 136. Springer, Berlin.

Ruthenberg H (1980): Farming systems in the tropics. Clarendon Press, Oxford (3rd ed.).

Rutherford MC (1980): Annual plant production-precipitation relations in arid and semi-arid regions. S Afr. J Sci. 76:53–56.

Sanchez PA, Logan TJ (1992): Myths and science about the chemistry and fertility of soils in the tropics. In: Lal and Sanchez, 35–46.

Sarmiento G (1984): The ecology of neotropical savannas. Havard Univ. Press, Cambridge, Mass. USA

Satoo T (1983): Temperature broad-leaved evergreen forests of Japan. In: Ovington, 169–189.

Schachtschabel P, Blume H-P, Brümmer G, Hartge KH, Schwertmann U (2002): Scheffer/Schachtschabel – Lehrbuch der Bodenkunde. Enke, Stuttgart (15th ed.).

Schmithüsen J (1968): Allgemeine Vegetationsgeographie. De Gruyter, Berlin (3rd ed.).

Schmithüsen J (ed) (1976): Atlas zur Biogeographie. Meyer, Mannheim.

Schnock G (1971): Le bilan de l'eau dans l'ecosysteme foret. Application a une chenaie melangee de haute Belgique. In: Duvigneaud, 41–47.

Scholes RJ (1990): The influence of soil fertility on the ecology of southern African dry savannas. J. Biogeogr. 17:415–419.

Scholes RJ, Walker BH (1993): An African savanna, synthesis of the Nylsvley study. Cambridge University Press, Cambridge.

Scholes RJ, Dalal R, Singer S (1994): Soil physics and fertility: the effects of water, temperature and texture. In: Woomer and Swift, 117–136.

Scholz F (1995): Nomadismus. Theorie and Wandel einer sozio-ökologischen Kulturweise. Erdkundl. Wissen 118. Franz Steiner Verlag, Stuttgart.

Schönwiese CD (2003): Klimatologie. Ulmer, Stuttgart (2nd ed.).

Schroeder D, Blum S (1992): Bodenkunde in Stichworten. Hirt, Kiel (5th ed.).

Schroeder F-G (1998): Lehrbuch der Pflanzengeographie. Quelle und Meyer, Wiesbaden.

Schubert R (1991): Lehrbuch der Ökologie. Fischer, Jena (3rd ed.).

Schulte A, Ruhiyat D (eds) (1998): Soils of tropical forest ecosystems. Springer, Berlin.

Schultz J (1984): Agrargeographie. In: Gaebe W et al. (eds): Sozial- und Wirtschaftsgeographie Bd. 3. Harms Handbuch der Geographie. List, Munich, 22–112.

Schultz J (1988): Die Ökozonen der Erde. Ulmer, Stuttgart. (3rd ed. 2002).

Schultz J (1995): Ökozonen. In: Kuttler W (ed): Handbuch zur Ökologie. Berlin (2nd ed.), 308–315.

Schultz J (1995): The ecozones of the world. Springer, Berlin.

Schultz J (1998): Ecozones, global. In: Meyers RA (ed): Encyclopedia of environmental analysis and remediation. John Wiley and Sons, Chichester, 1497–1518.

Schultz J (2000): Handbuch der Ökozonen. Ulmer, Stuttgart.

Schultz J (2000): Konzept einer ökozonalen Gliederung der Erde. Geogr. Rundschau 52:5–11.

Schultz J (2001 and 2002): Ökozonen der Erde. Peterm. Geogr. Mitt. 145/1–146/3.

Schunke E (1986): Periglazialformen und Morphodynamik im südlichen Jameson-Land, Ost-Grönland. Abh. Akad. Wiss. Göttingen 36. Vandenhoeck und Ruprecht, Göttingen.

Semmel A (1993): Grundzüge der Bodengeographie. Teubner, Stuttgart (3rd ed.).

Shaver GR, Billings WD, Chapin III FS, Giblin AE, Nadelhoffer KJ, Oechel WC, Rastetter EB (1992): Global change and the carbon balance of arctic ecosystems. BioScience 42:433–441.

Shugart HH, Leemans R, Bonan GB (eds) (1992): A systems analysis of the global boreal forest. Cambridge University Press, Cambridge.

Sick WD (2002): Agrargeographie. Das Geographische Seminar. Westermann, Braunschweig (3rd ed.).

Siegenthaler U, Sarmiento JL (1993): Atmospheric carbon dioxide and the ocean. Nature, 365:119–125.

Sims PL, Singh JS, Lauenroth WK (1978): The structure and function of ten western North American grasslands. J. Ecol. 66:251–285 and 547–597.

Sitte P, Ziegler H, Ehrendorfer F, Bresinsky A (1991, 1998): Strasburger – Lehrbuch der Botanik, Fischer, Stuttgart (33rd and 34th ed.).

Skartveit A, Ryden BE, Kärenlampi L (1975): Climate and hydrology of some Fennoscandian tundra ecosystems. In: Wielgolaski, Ecol. Studies 16:41–53.

Skujins J (cd) (1991): Semiarid lands and deserts – soil resource and reclamation. Marcel Dekker, New York.

Smith SD, Nobel PS (1986): Deserts. In: Baker NR, Long SP (eds): Photosynthesis in contrasting environments. Topics in Photosynthesis 7, Elsevier, Amsterdam, 13–62.

Solbrig OT, Medina E, Silva JF (eds) (1996): Biodiversity and savanna ecosystem processes: a global perspective. Ecol. Studies 121. Springer, Berlin.

Song Y (1995): On the global position of the evergreen broad-leaved forests of China. In: Box et al., 69–84.

Spaargaren OC, Deckers J (1998): The world reference base for soil resources. An introduction with special reference to soils of tropical forest ecosystems. In: Schulte and Ruhiyat, 21–28.

Specht RL (1981): Primary production in mediterranean-climate ecosystems regenerating after fire. In: Di Castri et al., 257–267.

Specht RL (ed) (1988): Mediterranean-type ecosystems: a data source book. Kluwer, Dordrecht.

Specht RL (1994): Species richness of vascular plants and vertebrates in relation to canopy productivity. In: Arianoutsou and Groves, 15–24.

Stäblein G (1987): Periglazial and Permafrost in Polargebieten. Münchener Geogr. Abh. Reihe B, 4. Munich, 97–107.

Stäblein G (1987): Periglaziale Mesoreliefformen und morphoklimatische Bedingungen im südlichen Jameson-Land, Ost-Grönland. Abh. Akad. Wiss. Göttingen 37. Vandenhoeck und Ruprecht, Göttingen.

Stoddart DR (1969): Climatic geomorphology: review and reassessment. Progr. Geogr. 1:160–222.

Sugden D (1982): Arctic and Antarctic. Blackwell, Oxford.

Sutton SL, Whitmore TC, Chadwick AC (eds) (1983): Tropical rainforest: ecology and management. Blackwell, Oxford.

Swank WT, Crossley DA Jr (eds) (1988): Forest hydrology and ecology at Coweeta, Ecol. Studies 66. Springer, Berlin.

Swift MJ (1979): Decomposition in terrestrial ecosystems. Blackwell, Oxford.

Swift MJ, Healey IN, Hibberd JK, Sykes JM, Bampoe V, Nesbitt ME (1976): The decomposition of branch-wood in the canopy and floor of a mixed deciduous woodland. Oecologia 26:139–149.

Tenhunen JD, Catarino FM, Lange OL, Oechel WC (eds) (1987): Plant response to stress function: analysis in mediterranean ecosystems. Springer, Berlin.

Thannheiser D (1994): Die Vegetationsvielfalt des kanadischen borealen Nadelwaldes. Essener Arb. 25. Schöningh, Paderborn, 1–21.

Thomas DSG (1988): The biogeomorphology of arid and semi-arid environments. In: Viles HA (ed): Biogeomorphology. Blackwell, Oxford, 193–221.

Thomas DSG (ed) (1989): Arid zone geomorphology. Bellhaven, London.

Thomas DSG, Middleton NJ (1995): Desertification: exploding the myth. John Wiley and Sons, Chichester.

Thomas MF (1994): Geomorphology in the tropics. John Wiley and Sons, Chichester (2nd ed.).

Tieszen LL (ed) (1978): Vegetation and production ecology of an Alaskan arctic tundra. Ecol. Studies 29. Springer, Berlin.

Tischler W (1990): Ökologie der Lebensräume. Fischer, Stuttgart.

Tischler W (1993): Einführung in die Ökologie. Fischer, Stuttgart (4th ed.).

Titlyanova AA, Bazilevich NI (1979): Semi-natural temperate meadows and pastures: nutrient cycling. In: Coupland, 170–180.

Tomaselli R (1981): Main physiognomic types and geographic distribution of shrub systems related to mediterranean climates. In: Di Castri et al., 95–106.

Tomlinson PB, Zimmermann MH (eds) (1978). Tropical trees as living systems. Proc. 4th Cabot Symp. Harvard Forest, Petersham (Mass.) 1976. Cambridge Univ. Press, Cambridge.

Tothill JC, Mott JJ (eds) (1985): Ecology and management of the world's savannas. Austr. Acad. Sci., Canberra.

Treter U (1993): Die borealen Waldländer. Das Geographische Seminar. Westermann, Braunschweig.

Tricar J, Cailleux A (1972): Introduction to climatic geomorphology. Longman, London.

Troll C, Paffen KH (1964): Karte der Jahreszeitenklimate der Erde. Erdkunde 18, 5–28.

Tuhkanen S (1984): A circumboreal system of climatic-phytogeographical regions. Acta. Bot Fennica 127:1–50.

UNEP (1983): Rain and stormwater harvesting in rural areas. Water Resources Series 5, Dublin.

UNESCO (1978): World water balance and water resources of the earth. UNESCO, Paris.1g

Unger PW, Sneed TV, Jordan WR, Jensen R (eds) (1988): Challenges in dryland agriculture. A global perspective. Proc. Intern. Conf. Dryland Farming Aug. 1988, Amarillo/Bushland, Texas. Texas Agricultural Experimental Station, Amarillo, Texas.

USDA Soil Survey Staff (1999): Keys to soil taxonomy, Pocahontas Press, Blacksburg, Virginia (3rd ed.).

Van Cleve K, Chapin III FS, Flanagan PW, Viereck LA, Dyrness CT (eds) (1986): Forest ecosystems in the Alaskan taiga. Ecol. Studies 57. Springer, Berlin.

Van Wambeke A (1992): Soils of the tropics. Properties and appraisal. McGraw-Hill, New York.

Venzke J-F (1990): Beiträge zur Geoökologie der borealen Landschaftszone. Geländeklimatologische und pedologische Studien in Nord-Schweden. Essener Geogr. Arb 21. Schöningh, Paderborn.

Vitousek PM, Aber JD, Howarth RW, Likens GE, Matson PA, Schindler DW, Schlesinger WH, Tilman GD (1997): Human alteration of the global nitrogen cycle: sources and consequences. Ecol. Applications 7:737–750.

Waelbroeck C (1993): Climate-soil processes in the presence of permafrost: a systems modelling approach. Ecol. Modelling 69:185–225.

Walker BH (ed) (1987): Determinants of tropical savannas. IUBS Monograph Ser. 3. IRL Press, Paris.

Walter H, Lieth H (1960–1967): Klimadiagramm-Weltatlas. Fischer, Jena.

Walter H, Breckle S-W (1983–1994): Ökolologie der Erde. Fischer, Stuttgart, 4 vols.

Walter H, Breckle S-W (1999): Vegetation and Klimazonen. Ulmer, Stuttgart (7th ed.).

Washburn AL (1979): Geocryology: a survey of periglacial processes and environments. Arnold, London.

Weischet W (1995): Einführung in die Allgemeine Klimatologie. Teubner, Stuttgart (6th ed.).

Weller G, Holmgren B (1974): The microclimates of the arctic tundra. J. Applied Meteorol. 13:854–862.

Werger MJA (ed) (1978): Biogeography and ecology of southern Africa. Dr. W. Junk, The Hague, 2 vols.

Werner PA (ed) (1991): Savanna ecology and management. Australian perspectives and intercontinental comparisons. Blackwell, Oxford.

West NE (ed) (1983): Temperate deserts and semi-deserts. Ecosystems of the World 5. Elsevier, Amsterdam.

Whitmore TC (1990): An introduction to tropical rain forests. Clarendon Press, Oxford.

Whittaker RH (1970): Communities and ecosystems. Macmillan, London.

Whittaker RH, Likens GE (1975): The biosphere and man. In: Lieth and Whittaker, 305–328.

Wickens GE (1998): Ecophysiology of economic plants in arid and semi-arid lands. Springer, Berlin.

Wielgolaski FE (ed) (1975): Fennoscandian tundra ecosystems. Ecol. Studies 16 and 17. Springer, Berlin.

Wielgolaski FE (ed) (1997): Polar and alpine tundra. Ecosystems of the World 3. Elsevier, Amsterdam.

Wirthmann A (1987): Geomorphologie der Tropen. Wiss. Buchgesellschaft, Darmstadt.

Woomer PL, Swift MJ (eds) (1994): The biological management of tropical soil fertility. John Wiley and Sons, Chichester.

World Atlas of Agriculture (1969): Instituto Geografico de Agostini, Novara.

Yamamoto S-J (1994): Gap regeneration in primary evergreen broad-leaved forests with or without a major canopy tree, Distylium racemosum, southwestern Japan: a comparative analysis. Ecol. Research 9:295–302.

Young MD, Solbrig OT (eds) (1993): The world's savannas – economic driving forces, ecological constraints and policy options for sustainable land use. Man and the Biosphere Ser. 12. UNESCO, Paris.

Zhong Zhang-Cheng (1988): Ecological study on evergreen broad-leaved forest. South West China Teachers University, Chongqing. (English summary).